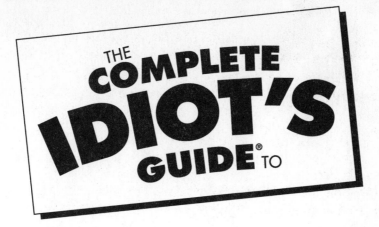

College Biology

by Emily Jane Willingham, Ph.D.

ALPHA

A member of Penguin Group (USA) Inc.

I dedicate this book to my family. All of them.

ALPHA BOOKS

Published by the Penguin Group

Penguin Group (USA) Inc., 375 Hudson Street, New York, New York 10014, USA

Penguin Group (Canada), 90 Eglinton Avenue East, Suite 700, Toronto, Ontario M4P 2Y3, Canada (a division of Pearson Penguin Canada Inc.)

Penguin Books Ltd., 80 Strand, London WC2R 0RL, England

Penguin Ireland, 25 St. Stephen's Green, Dublin 2, Ireland (a division of Penguin Books Ltd.)

Penguin Group (Australia), 250 Camberwell Road, Camberwell, Victoria 3124, Australia (a division of Pearson Australia Group Pty. Ltd.)

Penguin Books India Pvt. Ltd., 11 Community Centre, Panchsheel Park, New Delhi—110 017, India

Penguin Group (NZ), 67 Apollo Drive, Rosedale, North Shore, Auckland 1311, New Zealand (a division of Pearson New Zealand Ltd.)

Penguin Books (South Africa) (Pty.) Ltd., 24 Sturdee Avenue, Rosebank, Johannesburg 2196, South Africa

Penguin Books Ltd., Registered Offices: 80 Strand, London WC2R 0RL, England

Copyright © 2010 by Emily Jane Willingham, Ph.D.

International Standard Book Number: 978-1-59257-848-1
Library of Congress Catalog Card Number: 2009941614

12   11   10   8   7   6   5   4   3   2   1

Interpretation of the printing code: The rightmost number of the first series of numbers is the year of the book's printing; the rightmost number of the second series of numbers is the number of the book's printing. For example, a printing code of 10-1 shows that the first printing occurred in 2010.

Printed in the United States of America

Note: This publication contains the opinions and ideas of its author. It is intended to provide helpful and informative material on the subject matter covered. It is sold with the understanding that the author and publisher are not engaged in rendering professional services in the book. If the reader requires personal assistance or advice, a competent professional should be consulted.

The author and publisher specifically disclaim any responsibility for any liability, loss, or risk, personal or otherwise, which is incurred as a consequence, directly or indirectly, of the use and application of any of the contents of this book.

Most Alpha books are available at special quantity discounts for bulk purchases for sales promotions, premiums, fund-raising, or educational use. Special books, or book excerpts, can also be created to fit specific needs.

For details, write: Special Markets, Alpha Books, 375 Hudson Street, New York, NY 10014.

Publisher: *Marie Butler-Knight*
Associate Publisher: *Mike Sanders*
Senior Managing Editor: *Billy Fields*
Executive Editor: *Randy Ladenheim-Gil*
Development Editor: *Michael Thomas*
Senior Production Editor: *Janette Lynn*

Copy Editor: *Cate Schwenk*
Cover Designer: *Bill Thomas*
Book Designer: *Trina Wurst*
Indexer: *Tonya Heard*
Layout: *Brian Massey*
Proofreader: *Laura Caddell*

Contents at a Glance

Contents

13 DNA   153

14 Gene Expression   165

22 Systems 1: Homeostasis and Materials Exchange — 257

23 Systems 2: Regulation, Reception, and Response — 271

Appendixes

Introduction

Like species themselves, biologists come in all shapes and sizes. They don't all wear lab coats, they don't all grow beards and lose the hair on top of their heads. They don't all wear glasses. What they do have in common is a fascination with the living world. And they also often share a deep desire to transmit that fascination to others.

And that's why this book isn't just about studying biology or learning terms or memorizing facts. This book is a story, the story of life, how we study it, classify it, learn about it, and understand it in context. It's about satisfying our human curiosity about origins or how it is that certain things in life fit just right while others seem to be the product of a Rube Goldbergian deity laughing his head off somewhere. The great thing about the study of life is that with every little bit of knowledge we uncover, we open up a hundred new questions to answer. It's a never-ending quest that keeps the intellectually curious always wanting more.

But this book is also intended to help you. Every concept, every word or phrase is something I introduced because I know from experience what a reader of this book will need. I've emphasized what gets emphasized in college biology, and where I know there are knots for most students, I've sought to untie them for you, carefully and completely.

How This Book Is Organized

Part 1, "Life: The User's Manual," gets us started on the road to understanding life. There's a lot of chemistry here, and I spend a good deal of time building us up from atoms to molecules to the big molecules that make up living organisms. If you spend an equivalent proportion of time on truly understanding these concepts, what follows comes much more easily.

Part 2, "The Cell," is a pretty self-explanatory title. We tour the cell and learn about it as a factory for producing proteins. There's a (tiny) bit of physics as we learn where cells get the energy for all of that work (hint: from sugar). And then we learn about the organisms that build that sugar in the first place. Finally, a cell has to come from somewhere. That somewhere is another cell in the process of cellular reproduction.

Part 3, "Genetics," starts with a kind of cell division specialized for sexually reproducing species. (Yes, that means us.) Then we move on to Gregor Mendel, the brilliant monk-scientist who did intense work with pea plants. And finally, we get to the molecule that quite literally starts it all: DNA. It's your chance to unlock the secrets of how nature starts with DNA and ends up with you.

Part 4, "Evolution and Diversity," is a part with two words in the title that have powerful connotations beyond biology. But this is a biology book, and understanding evolution and the significance of diversity is crucial to understanding this science. They're also crucial to understanding how species arise, how we tell one species from another, and how we go about grouping species together based on their relatedness.

Part 5, "Form and Function (Plants and Animals)," focuses on these two groups of organisms. We start with plants—everything does anyway—and focus throughout on how nature shapes form in the context of the environment to fit function. This part also synthesizes some of the body systems in animals, showing, for example, how your respiratory, circulatory, digestive, and excretory systems work together to keep your body in balance. Finally, we cover the ways we develop from our one-cell beginning to the final product of trillions of cells.

Part 6, "Ecology," tackles why animals behave the way they do and zooms out from our focus on molecules and species to the big picture of communities, ecosystems, and the biosphere. This is the Big Picture part, the synthesis of everything that has come before.

Be sure to check out the website that accompanies this book at **collegebio.net** for a comprehensive glossary, more on microbes, a bio Q&A blog, and more.

The study of life has a few themes that thread through everything, from viruses to vampire bats:

◆ The overarching theme is evolution. As a famous biologist once said, nothing in biology makes sense unless it is placed in the context of evolution. Evolution is not presented here as a belief system but as something factual that confers order and sense on the millions of observations biologists have made over the centuries.

◆ Another theme of this book is cycles, the way energy cycles, the way life cycles, the way nutrients cycle. Nature recycles and reuses, and that theme is ever present here.

◆ A final overarching theme is interconnectedness. Again and again, I observe that no atom, no molecule, no gene, no organism, no system is an island. It's all connected to form the living, breathing global organism we call planet Earth.

Here and there, you'll find my opinion about something. As a biologist, I have a comfort zone for offering my own take on certain aspects of our studies of life. And you'll also find that I step into territory that many science teachers consider a No-Go Zone, and that is introducing humor and occasionally even anthropomorphism into the scientific explanations. I'm okay with that because I know that you, the reader, are far more likely to remember something you process while experiencing an emotion like amusement or anger. (I went with amusement, not being interested in making you mad.) You're processing the information, sure, but you're also linking it through your emotional processing center in your brain, which ensures better recall later on. How do I know that? Because I am a biologist. Read and enjoy this book. Maybe you'll choose to be one, too.

Extras

Throughout the book, I've included sidebars to coalesce core ideas, provide those eye-popping tidbits that only biology can offer, deliver a warning, or simply define some key terms.

Bio Basics

Key information or clarification of a "take-home" message.

Biohazard!

A warning to avoid common pitfalls.

def•i•ni•tion

Key terms in biology defined.

Bio Bits

Cool information or fascinating facts.

Acknowledgments

In a nod to genetics and environment, I start by thanking my parents, who bequeathed on me either genetically or environmentally or both a burning intellectual curiosity that only the study of life can satisfy. I also thank anyone and everyone who has ever told me that I can't do something—because of them, I have done many things I might otherwise never have thought of doing, including dropping everything to go to graduate school in science. My gratitude also to my siblings and my in-laws, whose excitement over this book may rival my own.

I'd like to extend my thanks to my agent, Bob Diforio, who patiently waited until I finally contacted him about something that really was a fit for me and then worked in his kindly, timely, and diligent way to facilitate success. Also owed my thanks for their help and patient support are Randy Ladenheim-Gil, executive editor at Alpha Books/Penguin Group, and Mike Thomas, "my" developmental editor.

Through my educational experience, I've had teachers who've influenced me for the good and for the bad, and either way, I'm grateful for the lessons I learned from them. In my experience as an educator, I've encountered just about every kind of student out there, and I owe them all a huge debt of gratitude: without them and their questions both lucid and loopy, I never would have had the tools to write a book like this. Thanks to them, I got to know them, and that helps me know who's reading this book.

Finally, as with many others who write acknowledgments, I've saved the best for last. I wouldn't be here or anywhere without the mutual love, support, and respect I have with my soul mate, partner, and husband, Marshall. And I wouldn't have the deep appreciation I do for the fascination biology can engender if I hadn't had the pleasure of engendering it in my own sons, Tom Henry, Will, and George. Without them, I'd just be a lonely biologist who writes stuff.

Trademarks

All terms mentioned in this book that are known to be or are suspected of being trademarks or service marks have been appropriately capitalized. Alpha Books and Penguin Group (USA) Inc. cannot attest to the accuracy of this information. Use of a term in this book should not be regarded as affecting the validity of any trademark or service mark.

Part 1

Life: The User's Manual

Life is built with atoms that form molecules that form bigger molecules that make up cells that make up organisms. Given the unavoidable involvement of atoms and molecules, we must unavoidably learn some chemistry. It's the chemistry of life, based on irresistible attractions without which you wouldn't be reading this book.

Life: We're Living It, Let's Study It!

In This Chapter

- ◆ Life defined
- ◆ Organizing life
- ◆ Organizing the interactions of life
- ◆ Evolution: the unifying process of life
- ◆ The scientific method
- ◆ Hypotheses, theories, and laws

You may have heard it before: *Bio* = "life" and *–ology* = "study of." And indeed, in biology, we study life. But that seven-letter word doesn't come close to illustrating the breadth and depth of that work. We focus a microscope on the smallest parts of a cell or rise above Earth's atmosphere and look at the entire living, breathing planet. The human body alone leads us to so many questions that hundreds of thousands of scientists all over the world focus on specific human systems and cells in their effort to understand more about us.

As the numbers of working biologists indicate, studying life is one of the greatest undertakings of humankind, one that for every answer yields more questions. One thing we do know: the smallest unit of life is the cell, and it is from a single cell that all life we see around us emerges.

Other entities, such as viruses, may share some of the features of life, but they do not have all of the properties that we have identified as "life" here on this planet.

How Do We Know Life When We See It?

Astrobiologists study life beyond Earth. You may, for example, have read about finding water on Mars, the first hint that life may have been possible there. But how will scientists know if they've actually found something *alive?* It won't be little green men but something microscopic, possibly very simple even at the molecular level.

We have several complicated ways of trying to determine indirectly if a distant planet carries life, but here on the home planet we have a straightforward checklist. Some properties of life as we know it on Earth include:

- **Reproduction.** People make more people, bacteria make more bacteria.

- **Energy use.** Life transforms energy. We grab the chemical energy in a molecule of sugar and use it to build molecules for ourselves.

- **Order.** Living things have order and structure. Just think of the complex structure of your wrist, ordered so that the bone, muscle, blood vessels, connective tissue, and skin are where they should be.

- **Response to the environment.** Poke a rock and it just sits there. Poke a sea anemone, and it retracts its tentacles.

- **Regulation.** A lizard moves to a sunny spot to warm up. We shiver with cold for the same reason.

- **Growth and development.** You began as a single cell. Look at you now!

- **Evolutionary adaptation.** Individuals best suited to their environment survive and reproduce, increasing those adaptive traits in the population. Bacteria do this when we use antibiotics against them: the resistant ones survive and the resistance trait becomes common.

Bio Bits
We once may have found—and killed—life on Mars. In the 1970s, Viking space probes may have picked up some microbes that emitted carbon dioxide. But the experiment fizzled when the Martian soil samples were exposed to water. We have since discovered earthly microbes with features for survival on Mars. They die when immersed in water. We may have found life in those Martian soil samples but promptly drowned it.

Thinking about these properties of life, let's consider the virus again. It consists typically of a protein coat wrapped around some genetic material. What properties of life does a virus have? Although they do evolve (all too well in some cases), reproduce (with help from cells), and exhibit order (very basic order), viruses lack some key features of life, including the ability to use energy on their own, exhibit growth and development, respond to the environment, and self-regulate.

Everything in Its Place: Organizing Life

The study of life is such an enormous undertaking that some scientists devote their careers strictly to its classification. In the fourth century B.C.E., Aristotle kicked things off with the idea of the *Scala Natura*, or Nature's Ladder. Along this ladder, also known as the "Great Chain of Being," all living things essentially strove to achieve perfection, with humans—specifically men—as the most perfect nondivine beings at the top and some kind of amorphous blob at the bottom. We've given up on climbing that ladder, having realized now that the cockroach, for example, is just as nicely suited to its environment as Brad Pitt is to his, if not more so.

The Tree of Life

Instead of a ladder, we've moved on to trees. You may have heard of the Tree of Life. The branching limbs of the tree reflect the idea that all life descends from all other life, with new branches emerging as changes accumulate. Think back on your own family history, with a great-great grandmother at the root, and the branchings that occurred to get to you. The Tree of Life is like that, except that it stretches back about 3.5 billion years (give or take a few hundred million) instead of a few generations. And instead of finding a great-great grandma at its root, you'll discover a basic, single cell parked there.

Of course, a lot has happened in the 3.5 billion years (again, give or take a few hundred million!) since the Grandmother of All Cells first divided to make new cells. In that time, hundreds of millions of species have arisen and gone extinct, and new ones arisen. We have no real idea of how many species there are, but put the number somewhere between 10 million and 100 million. As you can see, when you think big in biology, there's a lot of uncertainty built in. Life is uncertain, and the study of it can be, too.

The Hierarchy of Life

As humans, we're automatic classifiers. Being unable to categorize things leaves us agitated and confused. That's why we have a hierarchy of classification for organisms.

> **Bio Basics**
>
> A good way to remember the hierarchy is with this goofy but useful memory device: "King Philip Came Over For Ginger Snaps." The first initial of each word is the clue to the name of the hierarchical level.

It's like a file folder with several subfolders inside. The overall big folder is called a Domain. Inside the Domain folder, you'll find a Kingdom subfolder. Double-click on the Kingdom folder, and you'll find a folder labeled "Phylum." Every Phylum folder contains a folder called "Class." The other subfolders (becoming increasingly specific) are Order, which contains the Family folder, which contains the Genus folder, which contains the most specific folder of all, the Species folder.

You, as a human, are filed into the *Domain Eukarya*, based on our cell type. When that folder is opened, you find the folder *Kingdom Animalia*. We're in the Animal kingdom, along with all other animals, including insects, snails, and worms. Inside the Kingdom folder is our Phylum folder, labeled *Chordata*. It's a narrower selection than the Kingdom Animalia, and includes animals with backbones plus a few others.

Inside the Phylum folder is the Class folder. As you can see, our Class is *Mammalia*. This is an even narrower classification, including only animals that have fur and lactate. Inside our Class is our Order folder, labeled *Primate*. We'd be joined in this exclusive folder by lemurs, monkeys, and apes. Becoming even more exclusive, we go into the Family folder, containing the *Hominidae* family. Other members of this family are the gorillas, orangutans, and chimpanzees, our closest living relatives.

Clicking on the Family folder takes us to our Genus folder where, alas, we are alone as the only living representatives of the *Homo* genus. And of course, in our Genus folder, you find tucked away our Species folder, especially reserved only for *Homo sapiens*.

Themes and Leitmotifs: Organizing Biology

We need to organize not only organisms but also the structure and interactions of life. No organism is an island, and interactions begin even below the level of the cell, between molecules. So that's where we start when we classify living interactions.

Molecules and Organelles

A cell contains two sublevels of biological interaction. The first is the molecule. As we discuss in Chapter 4, life is built using four basic kinds of molecules. These molecules interact with each other through chemical bonds, laying the groundwork for life.

Cells also contain specific structures called *organelles*. Organelles are all made up of interacting molecules, and each organelle has specific duties to perform as the cell goes about its business.

def•i•ni•tion

An **organelle** is a specialized cellular structure that performs specific functions to support the cell.

Tissues, Organs, and Organ Systems

If a group of cells gets together and works as a team, executing the same function, we call that team a tissue. Any organism that is multicellular has tissues. Even a simple sponge has tissues specialized for different jobs.

Get a couple of tissue teams functioning together, and you've got an organ. Your outer covering, for example, has different tissues functioning together as the organ we call skin. Organs have specific functions to execute, such as "filtering" for the kidneys or "pumping blood" for the heart.

A bunch of organs working together toward a common goal is called an organ system. Your tongue, stomach, and intestines all work together for you to take in and absorb nutrients, and they form your digestive system.

Organisms, Populations, and Communities

Our next level of organization and interaction is the organism, any individual living thing. An organism can be as simple as a single cell made up of molecules and organelles, or as complex as a giant redwood tree, consisting of leaf tissues and root systems all interacting in the world's largest individual organism.

> **Bio Bits**
>
> Your body is home to a number of different bacterial communities. Recent research has found that a specific group of bacteria live on the human wrist, while a completely different collection resides in the crook of the elbow. And let's not even get started on what's living in your large intestine.

A forest of redwood trees is a population, all individuals of the same species living in the same area. Penguins huddled together against an Antarctic wind are a population, as are the same-strain bacteria colonizing the crook of your elbow.

The interaction of you and all the microbial critters living on and in you is called a community. A community is a group of different species that interact with each other in a complex web in a defined area. Communities include everything from the bacteria or fungi we can't see all the way up to the largest trees. Communities exist in the oceans, in forests, in deserts, on mountaintops, and deep in the earth's crust.

Ecosystems and the Biosphere

Remember those huddled penguins? They're also part of a community, a species interacting with other species—bacteria, fish, whales—in the unforgiving Antarctic environment. But these interactions aren't confined to living things; they also include interactions with the nonliving, or abiotic, environment. Rocks, climate, water, soil, light, atmosphere—all of these are abiotic factors that living things interact with in an ecosystem. There are many kinds of ecosystems, some of which seem incredible, including the communities interacting with their seemingly inhospitable abiotic environment inside the lava tubes of a volcano.

Add together all of the parts of the globe where life can be found—as deep as two miles into the earth's crust, in the mysterious darkness of an ocean abyss, and as high as the peak of Mt. Everest. These collected ecosystems of the earth form the biosphere, the largest unit of interaction, the living, breathing parts of our planet.

Evolution as the Common Thread

Examine the penguins, the inhabitants of the abyss, the redwood trees, and the bacteria on your arm, and you'll find evidence of the common thread that binds them all together: evolution. From the level of a molecule called DNA all the way up to visible common features—such as fur among mammals—evolution is the process that unifies all living things.

Evolution: Unifying Life on Earth

The child of two biological parents shares half of her genes with each parent. In turn, she shares a quarter of her genes with each of her biological grandparents. The percentage of identical gene sequences she shares with someone who is not a member of her family will be comparatively tiny. Take that short, two-generation scale and broaden it over millions of years, and the same relationship pattern emerges. The more distantly related we are to a species, the less alike our gene sequences will be. Like the child with her parents and grandparents, we are most genetically similar to our closest relatives.

Bio Bits
Theodosius Dobzhansky (1900–1975), an evolutionary geneticist, gave us one of the most famous—and truest—statements on evolution: "Nothing in biology makes sense except in the light of evolution."

In biology, we use these similarities to determine relationships among the millions of species that exist or have existed. We are filed away with chimpanzees into the same Family folder because they are our closest living relatives, with abundant visible commonalities and striking genetic similarities.

Evolution Is a Fact

So what is evolution? Evolution, in the biological sense, can be broadly defined as a change in a population over time. People may talk about evolution as a "theory," but it is, in reality, an established fact. Populations do change over time.

The theory that everyone talks about is related to *how* evolution happens, not *if* it happens. It's the difference between knowing that you have brown hair (a fact) and having a working scientific theory about how you ended up with it.

A familiar example of evolution may be the bacterial resistance described earlier. A population of pathogenic (disease-causing) bacteria multiplies inside of you and makes you sick. Some of the bacteria can resist an antibiotic called ciprofloxacin, while others succumb to it immediately. You weary of how the bacteria are making you feel, so you take some ciprofloxacin. The sensitive bacteria start dying off, but the resistant ones stick around.

Like many people, the minute you've reduced your bacterial load sufficiently, you start to feel better and forget to take your medicine. You've gotten rid of the sensitive bacteria in that population, but those resistant ones are now hale and hearty and ready to divide. They pass on their resistant genes to their daughter cells (in biology, cellular offspring are "daughters"), and lo! Now you have a population of bacteria that consists primarily of resistant organisms. That population has changed over time from a few resistant cells here and there to almost 100 percent resistant. That's evolution.

Bio Basics

Scientists often focus on the mechanism of how things happen. For example, many researchers investigate the mechanisms of cancer, or the pathways by which cells change to become cancerous.

As you can see, populations do change over time. We've made it happen ourselves, not only with bacteria but also in designing miniature horses, odd-looking goldfish, and dachshunds. We are an agent of that change over time, a *mechanism* for making evolution happen. Mechanisms, in science, are the pathways by which events take place. Humans, by specifically choosing dogs with very short legs and sausagelike bodies, have been a mechanism of change over time for the canine population.

There are several identified mechanisms by which evolution happens, but the one that gets the most press is natural selection. We cover that in detail in Chapter 15, but the point to take home here is that evolution itself is not the theory. The theory is *natural selection*, one possible mechanism for how evolution happens.

In brief, the theory is that nature selects the organisms that are best fit for the current environment to reproduce and pass along their "fit" genes. As more and more "fit" organisms accumulate in the population, the population may change over time.

How Scientists Study Life

Scientists can't just wander willy-nilly without some kind of plan in their quests to answer the questions of life. There is a process involved, known as the scientific method. The important thing to keep in mind is that this approach is a guideline, not a rigorous template that every scientist religiously follows for every investigation. We don't carry around "Scientific Method" reminder cards to ensure that we've covered every step. In the end, however, for anything to mature from a *hypothesis* to a theory or ultimately a law, each step usually must be satisfied many times.

def·i·ni·tion

A **hypothesis** is an educated but tentative answer to a thoughtful, well-framed scientific question. It must be testable. Also, because your hypothesis may be false, you have to be able to think of a test that would demonstrate either its validity or its flaws.

Method to the Madness: Hypothesis Testing

There are a few permutations of the scientific method, some with differing numbers of steps or maybe even a slightly different order. The following steps I give you flow logically from the process of science. Your instructor or textbook may have a slightly different pattern. And don't forget that this is simply a guideline. Any scientist who, for example, gets to Step 4 and realizes that his or her hypothesis is completely off base would simply start over.

1. **Observe:** You see a squirrel with a nut.

2. **Question:** Why does the squirrel have a nut?

3. **Hypothesize:** The squirrel is hungry and is going to eat the nut.

4. **Experiment, gather data:** Observe the squirrel and the outcome for the nut. Possibly add in other squirrels with other nuts to assess the outcome.

5. **Results:** The squirrel buries the nut instead of eating it. Other squirrels do the same.

6. **Conclusion:** Reject hypothesis as wrong. New question! Why did the squirrel bury the nut?

Lather, rinse, repeat. Science is a never-ending quest for knowledge, always asking testable questions based on observations of the world around us. You've been on this quest your entire existence and likely have followed many of the previous steps in your everyday life. For example, you realize you've lost your cell phone just as you leave a restaurant. You observe the loss (Step 1) and think, *Now where did I put that thing?* (Step 2). You hypothesize that you left it on the table in the restaurant you just left (Step 3). This hypothesis is completely testable and falsifiable: All you have to do is gather data (Step 4) by returning to the table to see if the phone is there. You check the table and find your phone (results are in! Step 5). Your conclusion? Your hypothesis was correct and you do not reject it (Step 6. Done!). See? You're a scientist, too!

Theories and Laws

Sometimes, a hypothesis undergoes the process described earlier over and over again, and every time scientists get to Step 6, the hypothesis cannot be rejected—it still stands. When a hypothesis has taken a beating like that and remains upright, it may be ready to mature and broaden into a theory. This is especially true if the hypothesis is still standing while alternative hypotheses have fallen all around it, defeated.

A theory in science has undergone an intellectual and experimental pounding and proven itself to be generalizable. Designating something as a theory in science is not a casual process, as when someone says, "My theory is that the sauerkraut isn't selling because people just don't like cabbage." Designating a scientific concept as a theory requires a rigorous process—usually many people experimenting, testing, publishing, and having their work critiqued. It might take years, decades, or even centuries before a theory is established. Often, several mutually supportive hypotheses that have withstood the test of time collate into a broader concept that we think of as a scientific theory. It's a broad explanation that covers many things, rather than a small, specific explanation for a single kind of observation.

> **Biohazard!**
>
> In science, scientifically speaking, we never say that something is "just" a theory. A theory has withstood multiple challenges and deserves enormous respect.

A natural law is a phenomenon that happens the same way every time under certain conditions. Here on Earth, for example, if you step out of a helicopter without a parachute you will fall downward rather than upward. There is a law that describes this invariable outcome: things with mass are attracted to each other, and bigger masses hold bigger attractions. Thus, when you step from the helicopter, the huge mass of the earth attracts your much smaller mass toward it. Unless someone turns off the gravity switch, that law is expected to hold.

The Least You Need to Know

- Life has specific properties, including the ability to reproduce, evolve, grow, use energy, and respond to the environment.

- The cell is the smallest unit of life with these properties.

- Living things are classified into a hierarchy of the broadest to the most specific, and organized into levels of interaction.

- Evolution, which unites all life on Earth, is change in a population over time, and natural selection is one theorized mechanism of evolution.

- The scientific method is a rational guideline, not a rigorously followed checklist.

- A hypothesis is a testable scientific answer to a question, while a theory is broader and has withstood significant testing, and a natural law is invariable.

Chemistry: Irresistible Attractions of Life

In This Chapter

- Atoms and subatomic particles
- Electronegativity
- Forming molecules
- Intramolecular bonds
- Intermolecular bonding
- Water as the solvent of life

I know, I know. This is a biology book. Why a chapter on chemistry? Well, the fact is that you can't really understand anything about biology unless you understand the molecules that make up living things, and you can't understand molecules and how they interact unless you understand … chemistry. Don't worry. This is chemistry at its most shallow, need-to-know basis—just enough so the rest of this book makes biological sense.

The Basics: Atoms and Molecules

Atoms and molecules form the intersection of physics, chemistry, and biology. The atom is the realm of physicists, but without interactions between atoms, we wouldn't have chemistry, and without the outcome of these interactions—molecules—we wouldn't have biology. In fact, as you will learn in Chapter 3, we can even break the millions of molecules of life down into four basic categories. But to understand those categories, we first need to understand where molecules come from.

Atoms: Without Them, We're Nothing

What is an atom? The atom is the smallest unit of an element, which is any substance consisting of the same kind of atom. Chemists cannot break atoms apart, although physicists, who have a few tricks up their sleeves, can. Reduced to basics, an atom consists of a center or nucleus containing *protons* and *neutrons*. Around that nucleus, attracted to its positively charged protons, are the *electrons*, relatively tiny, negatively charged particles.

def•i•ni•tion

A **proton** is a positively charged particle in the nucleus of an atom. A **neutron** is a particle with no charge, also found in the nucleus. An **electron** is a negatively charged particle that orbits the atomic nucleus. Electrons are about $\frac{1}{1800}$th the mass of protons or neutrons, which have roughly equal masses.

The number of protons in an atom is its atomic number and determines its identity. Atoms with the same atomic number are all atoms of the same element. Thus, all atoms with one proton are hydrogen atoms, and all atoms with two protons are helium atoms.

In chemistry as occasionally in life, opposites attract. The number of protons in an atom determines how many electrons it has. Thus, a hydrogen atom, with its single proton in the nucleus, attracts a single electron (hydrogen has no neutrons). In terms of content, hydrogen is the simplest atom and can even wander around as a lone proton, without the electron in attendance.

Electrons don't gather around the positively charged nucleus like a crowd rushing the stage at a show; in fact, they don't even get that close to it, atomically speaking. They must order themselves into specific, whole-number shells around the atomic center. The first shell, nearest the nucleus, is front-row seating with room for only two electrons. The second and third shells can each hold eight electrons. Shells and their capacities become more complex beyond this level, but for our purposes, three shells out is plenty.

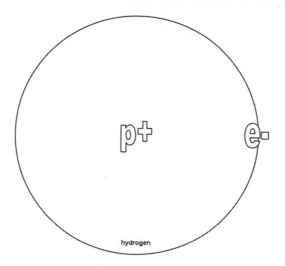

Hydrogen contains a proton and an electron but no neutrons.

Neutrons don't have a charge (hence their name, which suggests neutrality), but they do have mass. Their presence adds heft to the atom but doesn't affect the atom's interactions with other atoms. You can think of an atom that packs on a couple of neutrons as still being the same atom, just a little bit heavier.

Isotopes: More to Love

A good example of the effect of extra neutrons on atoms is carbon. Carbon has six protons and six neutrons in its most common state, carbon-12 (6 protons + 6 neutrons = 12). This gives it an atomic mass of 12. It can add on a couple of neutrons, resulting in carbon-14 (6 protons + 8 neutrons = 14, for an atomic mass of 14). This carbon isotope—or different form of the atom—still has six protons, so it's still carbon. With the extra neutrons in carbon-14, there's just more of the carbon atom to love. While the naturally occurring isotopes of carbon have more than 12 neutrons, it is possible for isotopes to have fewer neutrons, too.

Bio Bits
Most of the earth's relatively limited amount of carbon-14 is produced in the atmosphere. Plants take it up, so the relative amount of atmospheric carbon-14 at a specific time will match the relative amount in a plant or an animal that eats the plant. The difference between atmospheric carbon-14 levels from a specific time point and levels in a sample tell us the age of the sample. Carbon dating is accurate to 60,000 years.

Carbon-14 still has the same number of electrons as carbon-12, so its chemical behavior remains essentially unchanged. The number of electrons available in the atom's outer shell determines an atom's chemical behavior, or how an atom interacts with another atom. This number of outer electrons is known as the atom's valence.

Valence: A Quest for Stability

Of the six electrons of carbon, two take their places in the first shell, leaving four more to occupy the second shell. Because that second shell can hold eight electrons, carbon atoms have four empty electron spaces in their outermost shells. Thus, carbon has a valence of 4.

Atoms with the same number of protons and electrons are neutral in terms of charge. But another factor leaves many atoms looking for atomic buddies for bonding: stability. Atoms are considered stable when their outer electron shells are full to capacity.

Bio Basics

Most elements on the periodic table are neutral as lone atoms, with equal numbers of protons and electrons, but unstable in terms of valence. When they form bonds, they become stable but lose their neutrality. The exception is the noble gases, which have the same numbers of protons and electrons and have full outer shells. Thus, as lone atoms, they are neutral and stable.

In pursuit of stability, atoms bond with other atoms. Risking the ire of chemistry teachers everywhere, I am going to anthropomorphize atoms here, describing them as "wanting" and "looking for" other atoms. Carbon, lacking four electrons in its outer shell, is "looking for" four more electrons to have a feeling of full, satisfied outer shell stability. When atoms engage in this electron sharing or transfer, they form bonds, and in forming bonds, they make what we call molecules. They're also engaging in what we call chemistry.

Molecules: The Result of Chemistry

Imagine a lone carbon atom, wandering around, its outer shell feeling like something's missing. And there is: it has four empty electron spots. If carbon could just find four more electrons, it could relax into a relationship with another atom or atoms, fulfilling one another's valences for life.

And in a way, minus the soap-operatic anthropomorphizing, that's what carbon and any other atom with space in the outer shell do. They find other atoms with a need—either to share an electron, give up an electron, or take an electron—and interact with those other atoms, forming bonds. As you can see, chemistry is not that different from the dating scene.

For many elements, the periodic table is a quick way to tell how many electrons are in the outer shell and how many the atom needs to be stable. Elements in the same column have the same valence, regardless of their total electron number. Thus, the elements in the first column of the periodic table—including hydrogen, sodium, lithium, and potassium—all have one electron in their outer shells. The elements in the second column all have two. Over on the right of the table, all of the elements in column seven need just one electron to feel full.

Bonds, Atomic Bonds

Most atoms can form a bond with some other atom. The exceptions are the *noble gases*. These atoms in the final, far right column all have outer shells with every electron space filled. They have no need for chemical interactions, and we refer to them as "inert." They're perfectly satisfied, thank you very much.

The commoner atoms—those lacking the nobility of a full-shelled noble gas—must seek their fulfillment through other atoms. This fulfillment can come in a few forms, all dependent on how "grabby" a given atom is with another atom's electrons.

Electronegativity: Not Sharing Nicely

The "grabbiness" of an atom for an electron is called its electronegativity. Some atoms are like two-year-olds—they just don't understand how to share properly.

Atoms vary in their grabbiness. Some have a weak pull on electrons, with electronegativities less than 1, while others have electronegativities near 4, the peak of grabbiness. The grabbiest atom of all is fluorine. If a hapless sodium atom (electronegativity = 0.93), with a sole electron lurking in its outer orbital, gets near a fluorine, sodium can kiss that electron goodbye. Fluorine, an electron-grabbing bully with an electronegativity of 3.98, will practically strip sodium's electron right off.

It's like having someone 4.28 times your strength take away your beach ball. Strangely enough, once fluorine has ripped away sodium's electron, sodium will have a full outer shell and a feeling of peaceful stability, even though an electronegative bully just took its electron.

Some atoms are grabby with electrons but not so grabby that they can completely take the electron for themselves. Sure, they'll share, but not nicely. Oxygen is a good example. It's pretty grabby with electrons (electronegativity = 3.44) but will share with hydrogen (electronegativity = 2.20). Yet even as oxygen shares, it pulls the electrons slightly closer to itself and slightly away from weaker hydrogen. This difference yields some remarkable results that we'll discuss in the next chapter as we discover why water is weird.

Spare an Electron? Ions and Ionic Bonds

When electronegativities between two atoms are extremely imbalanced, their interaction results in another imbalance. The imbalance in this case is the number of negative charges each atom will have once the electron exchange is complete.

A very imbalanced pairing is the sodium-fluorine interaction described earlier. Fluorine on its own has exactly enough electrons (9 of them) to balance out its number of protons (9). Rudely grabbing some other atom's electron gives fluorine 10 electrons and only 9 protons.

Fluorine has now ionized and has an extra negative charge. When an atom has a charge imbalance, it becomes an ion. Fluorine, in grabbing sodium's sole outer-shell electron, becomes a negative ion, Fl-. This loss leaves sodium with 11 protons but only 10 electrons. Sodium is now a positively charged ion, Na+. The two atoms have formed an *ionic bond*.

As another example, let's take the famed combination of sodium and chloride. Sodium has such a low electronegativity that it will give up its sole outer electron to just about anything, so chlorine (electronegativity = 3.16) has no problem grabbing that electron. In doing so, chlorine becomes the *anion* Cl-, leaving sodium as the *cation* Na+. When they are in their ionic bond interaction, we refer to them as NaCl, or sodium chloride—common table salt.

> **Biohazard!**
>
> Don't make the mistake of thinking that electrons transfer completely in an ionic bond. It's never a complete transfer, even in the most imbalanced interactions.

> # def•i•ni•tion
>
> An **ionic bond** can form between two oppositely charged ions. A negatively charged atom is called an **anion**; one with a positive charge is a **cation**.

You Complete Me: Covalent Bonds

Not all interacting atoms are so out of balance in electronegativity. If you recall hydrogen and oxygen, they will share an electron pair, rather than ionizing. When atoms have electronegativities similar enough for sharing, rather than ionizing, we call the bonds they form covalent bonds (*co* = sharing, and *valent* = electrons in outer shell).

If the electronegativities are very similar, the sharing will be roughly equal in a bond we call a nonpolar covalent bond. A good example is the bond between two hydrogen atoms to make H_2, or hydrogen gas. Their sharing is exactly equal because their electronegativities are exactly the same, so the bond is nonpolar. Identical atoms always have the same electronegativities.

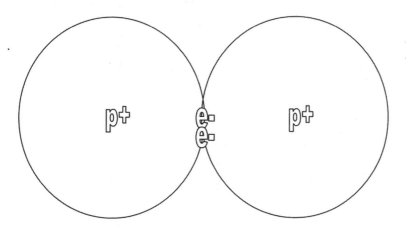

Two hydrogen atoms equally share an electron pair in a single covalent bond.

hydrogen gas

The bonding can be more complex than two atoms sharing an electron pair. Carbon and hydrogen are a good example. They have very similar electronegativities (2.55 versus 2.20). Carbon has four outer-shell electrons, but a hydrogen atom needs only one to complete its shell. The chemistry deities can stick four hydrogens on a single carbon atom, each in a single covalent bond with carbon. The resulting molecule has four hydrogen spokes around the central carbon. This molecule, with its four single covalent bonds, is CH_4, or methane.

Biohazard!

Avoid the trap of thinking that "single" refers to sharing only one electron. A single covalent bond refers to sharing an electron *pair* between two atoms, one from each atom. A double covalent bond involves the sharing of two pairs of electrons, or a total of *four*.

Methane consists of a single carbon involved in four single covalent bonds respectively with four hydrogens. The sharing relationship in each bond is roughly equal.

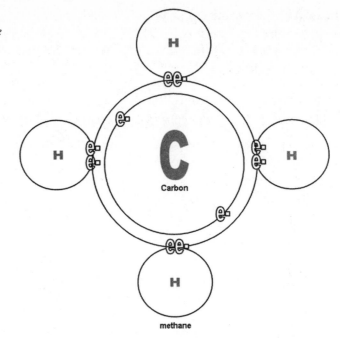

methane

There can also be double covalent bonds, as happens when two oxygen atoms bond to form oxygen gas, O_2. This sharing relationship is completely balanced in terms of electronegativity because the two atoms involved are the same.

A molecule of oxygen gas consists of two oxygen atoms equally sharing two pairs of electrons in two covalent bonds to form a double covalent bond.

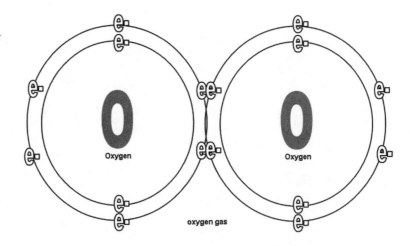

oxygen gas

So far, we've talked about atoms bonding together to form molecules, either with ionic or covalent bonds. But molecules themselves can bond with one another in intermolecular bonds, which has everything to do with the electronegativities of the atoms in each molecule.

Molecular Bonds, Inside and Out

The interactions that build a single molecule are called intramolecular bonds (*intra* = within). The two kinds of bonds we've discussed so far, ionic and covalent, are intramolecular bonds. But often, a molecule will interact, at least transiently, with another, separate molecule.

Intramolecular vs. Intermolecular

The big molecules of biology, built through intramolecular bonding, also can interact with each other via a type of weaker, intermolecular (*inter* = among or between) bond. This weaker form of bonding has a potentially misleading name: hydrogen bond. It got its name from the behaviors of water molecules, which interact because the hydrogens that are bonded with the oxygen in the molecule have the weaker electronegativity. The result is a grabby oxygen in the water molecule sharing electrons with the hydrogens but pulling that negative charge closer to itself, away from the hydrogens.

A water molecule thus has a couple of slightly negative areas on the oxygen side and a slightly positive charge around each of its hydrogens, as the figure on the following page shows. These charged areas are not as fully negative or positive as we find with ions. Because of these two distinct ends, or poles, of the water molecule, we say that it is polar. The covalent bonds holding the oxygen and hydrogens together are thus polar covalent bonds because of the unequal sharing of electrons.

 **Bio Basics**

Atomic bonding is a continuum. The relative electronegativities between the atoms involved determine the differences in bonds. Ordered from the most extreme differences to the most equal sharing, the order would be ionic→polar covalent→nonpolar covalent.

When a slightly negative oxygen area of one water molecule is near a slightly positive hydrogen area of another, the two molecules form the relatively weak intermolecular hydrogen bond. In a sample of liquid water, these molecules are constantly forming and breaking and reforming these weak bonds.

On the left is a water molecule, with the positive and negative portions indicated. On the right are four water molecules interacting with a fifth, at center, through hydrogen bonds (dotted lines).

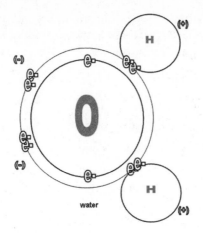

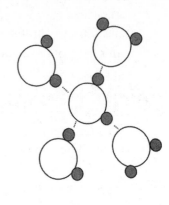

The word "slightly" makes this scenario seem almost unimportant. After all, why would being slightly charged be such a big deal? Well, these "slightly" charged areas of a water molecule form the basis of life on Earth.

Biohazard!

Don't think that the "hydrogen" in "hydrogen bond" means that any molecule with hydrogen can form a hydrogen bond. For example, in methane, or CH_4, the C and H electronegativities are equal enough that the molecule is not polar and doesn't form hydrogen bonds.

What Can Water Dissolve and Why?

Have you ever wondered why sugar or salt will dissolve in water, but butter does not? Here you will find the answers to these mysteries and more.

A sugar cube you drop into water is probably sucrose and consists of carbon, hydrogen, and oxygen. Of these, carbon and hydrogen have roughly similar electronegativities, with no areas of slightly negative or positive charge. But those few big fat oxygens in that sugar are grabby. Anywhere oxygen shares electrons with hydrogens, that sugar molecule is going to have an area of slight charge. Because the orientation is with the hydrogen to the outside, the sugar faces the world with a few areas of slightly positive charge.

Water molecules can orient themselves around a molecule of sugar so that their areas of slight negativity can hydrogen bond with the areas of slight positivity on the sugar molecule, as the following figure shows. They basically surround the molecule, bonding to these slightly charged areas, and extract it from its sugary cohort. This is how sugar dissolves in water.

Salt, like NaCl, is similar. Water molecules orient their negative oxygens to the Na+ and extract it, and they orient their positively charged hydrogens to the Cl- and extract those, as in the following figure. In this way, with water surrounding its separate ions, salt dissolves in water.

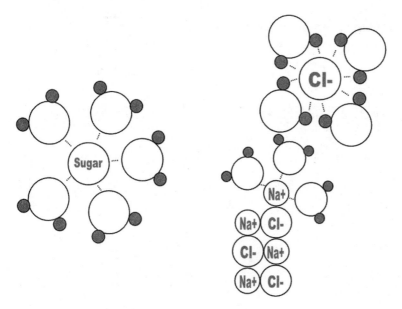

Because of unequal electron sharing between oxygen and hydrogen, water is a polar molecule that can hydrogen bond with other polar molecules, like sugar and salt.

The imbalanced electronegativities of these molecules allow them to interact via the weak, intermolecular hydrogen bonds. Butter, which consists mainly of carbon and hydrogen (as all fats do), doesn't have this imbalanced electronegativity—carbon and hydrogen share pretty equally, and there are no areas of slight charge. So a pat of butter in water just sits there. Water molecules find nothing attractive in the butter. A drop of water on butter holds its droplet form, more attracted to itself than the butter.

Now you know why oil and water don't mix. Oil, a fat like butter, is nonpolar and doesn't interact with the polar water molecules. In fact, oil and water repel one another. This repulsion is the basis of waterproof clothing. A plastic raincoat, for example, is made from oil (a fat) and thus repels water.

The ability of water to hydrogen bond with so many substances (except fats) and dissolve them has led to its designation as the "solvent of life." (A solvent is a liquid in which substances are dissolved.) But the ability of water to hydrogen bond with itself gives it a special property that makes it a fundamental requirement for life itself, as we'll see in Chapter 3.

The Least You Need to Know

- An atom's chemical behavior or interaction with other atoms stems from its valence.

- Interacting atoms with very different electronegativities can ionize and form ionic bonds.

- Covalent bonds result when atoms with relatively similar electronegativities share electrons and form a molecule.

- When covalently bonded atoms share electrons unequally, the molecule can be polar, with positive and negative areas.

- Polar molecules containing hydrogen can form intermolecular bonds with each other called hydrogen bonds.

- Because of its polarity, water can dissolve substances like sugar and is considered the "solvent of life."

Water and Carbon, a Recipe for Life

In This Chapter

- Water as the basis for life on Earth
- Water behavior
- The pH scale
- The elements required for life
- Introducing carbon and functional groups

Water is weird, and thanks to its odd properties, we're all here to learn about it. Probably the strangest thing about life on Earth is the fact that we can distill all of the material explanations for existence down to a single factor: the differential electronegativities of hydrogen and oxygen. Without their unequal sharing partnership, water wouldn't do what it does, life on Earth likely wouldn't have been able to arise or at least thrive, trees wouldn't be able to stretch to the skies, and our home planet would be a virtually unrecognizable wasteland of alternating extreme cold and heat. Kind of makes you appreciate water a bit more, doesn't it?

And then there's carbon, the other ingredient for life. Thanks to its special versatility, nature has used it as the centerpiece of the molecules of life. These two ingredients, water and carbon, in the recipe of life are the fundamental reasons we can be here right now, learning about the life they build and support.

Water Is Weird

For most substances, the densest state is solid, and if you drop the solid form of something into its liquid form, the solid will sink. Think of dropping a cold stick of butter into a vat of melted butter. You fully expect the solid stick of butter to sink, and it does.

Water does not behave this way. If you drop a cube of ice—the solid form of water—into a glass of liquid water, the ice floats. Why does solid water float on liquid water? The answer is hydrogen bonds.

When water molecules form hydrogen bonds with each other, if there's sufficient heat present, these bonds break and reform easily. Thus, in liquid water, water molecules are pretty closely packed as the bonds between them break and molecules bump into and bounce off of one another. It's like a bustling, jostling crowd at a fair, touching and moving apart, over and over.

Right at about 4°C (in science, everything is in Celsius!), the molecules still move just enough to break bonds, but not so much that they bounce away a great distance. Think of a circle of people, all holding hands while also keeping their arms close to their sides, swinging them slightly. Like water molecules bonded at 4°C, they'll be tightly packed. At this temperature, water molecules are at their closest.

As the temperature decreases below 4°C, though, the molecules slow down even more. The bonds between them stiffen, pushing the molecules apart. Imagine our circle of people again, but now spreading their arms out to the side as far as possible while still holding hands. Our tight little circle will expand into a larger circle. Still the same number of people, just occupying more space. That's what happens to water between 4°C and 0°C. The hydrogen bonds are stuck by the lack of heat, pushing the water molecules far apart from one another.

> **Bio Bits**
>
> Water reaches peak density at 4°C.

When you apply heat to cold butter, you expect that butter to expand as it melts. If you put the melted butter into the freezer, you expect it to contract as the heat leaves the sample. Water doesn't act like that. Water, unlike most other substances, *expands* as the temperature cools from 4°C to 0°C (freezing). As heat leaves the sample, the *kinetic energy* of the water molecules decreases, the hydrogen bonds stiffen up, and the molecules move apart from one another, expanding the space that they occupy.

Thus, in the solid form (ice) at 0°C, water molecules are stuck in hydrogen bonds, forced into a position they can't escape because of their lower energy. At about 0°C, the bonds freeze the molecules apart from one another in a crystalline lattice. At this temperature, there is more space between the molecules than when water is liquid at 4°C. So the less-dense solid (ice) will float on the denser liquid. Not much else behaves like this, and this behavior of water is a key to the development of life on Earth.

Why Water = Life

It's no accident that Earth is about two-thirds water and so are we. For billions of years, the blue planet has experienced periods locked in ice, followed by periods of thaw. Now imagine if ice sank rather than floated on liquid water. Every time our planet froze over, the ice would sink to the bottom of the ocean, locking the earth's core in a frozen vault.

Keep imagining that this happens a million times over billions of years. Eventually, Earth isn't a blue, watery planet but an inhospitable block of ice.

But the earth isn't like that. Thanks to the fact that ice floats, it's available to the heat for thawing, it blankets the liquid deeps beneath it, allowing life to flourish, and our liquid planet provides the perfect breeding ground for the molecules of life to meet, interact, and eventually yield constructs that bear the recognizable features of life.

Cohesion, Adhesion, Modulating Temp

In water, molecules can meet and interact, creating the enormous macromolecules that form cells and their structures. But water also helps modulate temperature. It has a relatively large capacity to take up heat without changing temperature, a feature we owe in part to the hydrogen bonds and the high *specific heat* of hydrogen.

Heat must be great enough to overcome these bonds and increase the molecule's

def•i•ni•tion

Specific heat is the amount of heat required to raise the temperature (average kinetic energy) of 1 g of a substance by 1°C. Water sets the bar for specific heat at 1. Anything with a lower specific heat requires less heat energy than water to raise 1 g of it a degree.

Bio Basics

It takes a lot of heat to get water molecules moving enough to increase their kinetic energy enough to increase temperature. To put this into perspective, the specific heat of iron is about a tenth that of water. If you put a liter of water on high heat in an iron skillet, which one would you dare to touch after about a minute of applying the high heat?

kinetic energy before the temperature will increase. Because water can take up so much heat before changing temperature, it helps modify the temperature of what's around it. Without water, the earth would be a rocky desert of temperature extremes, not quite as bad as Venus, but almost.

You can realize the importance of high specific heat by checking the difference in average temperatures between coastal and inland areas at the same latitude. San Francisco, which is on the northern U.S. West Coast, lies at a latitude of 37.775. Brushy, Illinois also lies at that latitude, but inland. San Francisco's average high temperatures range from 58°F in winter to 71°F in summer, a difference of 13 degrees. The average highs around Brushy range from 40°F in the winter to 89°F in the summer, a differential of 49 degrees. As you can see, water has a smoothing effect on temperatures.

def•i•ni•tion

Cohesion is the ability of water molecules to stick to each other through hydrogen bonds. **Adhesion** is the ability of water molecules to interact with something else, such as the surface lining the insides of the vessels a plant uses to transport water.

We owe the trees above our heads to two other properties of water, *cohesion* and *adhesion*. Because of their ability to interact with other molecules through intermolecular bonds, water molecules can stick onto surfaces while sticking onto each other. This ability is especially useful to trees.

Leaves exposed to sunlight receive heat radiation. The heat can agitate water molecules just under the leaf surface so much that they leap free of the leaf. These water molecules escape, or evaporate. Their energetic exit while still bonded to other molecules tugs the other water molecules in the tree upward against gravity. The molecules exhibit cohesion with one another as evaporation pulls them upward.

Those cohesive water molecules being pulled upward by evaporation are also adhering to the inside of the tree. The plant has tubes specifically designed to carry water throughout the plant, just as your blood vessels are designed to carry your blood. Water molecules can interact with the lining of these water-carrying tubes, exhibiting adhesion to this nonwater substance. Obviously, the lining of these tubes must be polar to adhere with water in this way.

pH: The Basics (Acids and Neutrals, Too)

Water molecules can break up into ions, yielding hydrogen ions (H+) and hydroxide ions (OH-). The balance between these two kinds of ions in a solution determines whether or not that solution is acidic or basic. If the balance tilts to a majority of H+, the solution is acidic. If it tilts to OH-, the solution is basic. If it's about even, the solution is neutral.

Here is a list of acids: HCl, H_2SO_4, HNO_3. You may notice something that they all have in common: they all contain H. The reason they're acids is that when any one of these compounds is mixed with water, it will release its H as an ion, H+, into the solution. Anything that increases H+ concentrations in a solution is called an acid.

In fact, we base the entire pH scale on the concentration of H+ in a solution. If the H+ concentration is 1×10^{-3} M (M, meaning molarity, is a unit of concentration that indicates indirectly how many grams of a substance there are per liter of liquid), that's a relatively high concentration of H+ and a pretty acidic solution. Vinegar, for example, has this acidity.

But when we're talking about pH, we don't give the H+ concentration. We take a shortcut. If the exponent of the concentration is $^{-3}$, the negative log of that is 3. To avoid writing out lengthy exponential equations, instead we take the negative log of the concentration—in this case, 3—and designate that as the solution's pH.

Bio Basics

Some biology classes get into details about calculating concentration or molarity. We're sticking with the simple explanation here that concentration refers to the grams (or other unit of mass) of substance per volume of solution it's dissolved in. If a solution of sucrose and water has 342 g of sucrose per liter, its concentration is 342 g/L. If it has 684 g per liter, the concentration of sucrose in the solution is doubled, or twofold the first one.

We use the log scale because if we stuck with the actual concentrations, the scale would span changes that could be 100-trillionfold or more. That can get kind of awkward for scientists who have to describe a range of pH values for experiments.

Thus, for a solution with a very low concentration of H+ at 1×10^{-10} M, the pH is 10. Notice that because the exponent is negative, as the absolute exponent value goes up, the concentration gets smaller. A solution that is 1×10^{-4} M H+ has more hydrogen ions than a solution that is 1×10^{-5} M. These solutions are, respectively, pH 4 and pH 5.

The pH scale itself spans from 0 to 14. The midrange is pH 7, which is considered neutral, an even balance of H+ and its counterpart in pH, OH-.

Biohazard! _____

When mixing solutions using acids (or bases, for that matter), never put your acid or base into a container first; instead, "Do what you oughta. Add the acid (or base) to water."

It's easy to determine the OH- concentration of a solution if you know the pH. If the pH is 2, you know that the H+ concentration is 1×10^{-2} M. The pH scale ends at 14. To determine the exponent for the OH- concentration, you subtract exponents: $14 - 2 = 12$. So that leaves the OH- concentration as 1×10^{-12}. As you can see, the concentration of OH- is quite low (0.000000000001) compared to that of H+ (0.01) at this very acidic pH.

Just as donating an H+ to a solution identifies an acid, donating an OH- to a solution is the hallmark of a base. Basic solutions have a pH higher than 7 and up to 14, a low H+ concentration, and a high OH- concentration.

The pH scale and some common solutions with specific pH values.

	pH	Common solutions with that pH
Basic	13–14	Lye, bleach, oven cleaner
	12	Soapy water!
	10–11	Household ammonia (11.9), milk of magnesia (10.5)
	9	Toothpaste
Neutral	7–8	Seawater, baking soda, tap water
Acidic	6	Milk, urine
	5	Black coffee
	4	Tomato juice
	3	Soft drinks, orange juice
	1–2	Stomach acid (1), lemon juice (2.3)
	0	Battery acid

This discussion of pH sounds largely chemical, doesn't it? But pH matters in biology. For every organism, there is usually an optimal pH. Often, different parts of that organism can have different optimal pH values. Cells in our blood, for example, do well in a pH around 7.4, but the enzymes in our stomach need a burningly acidic pH of 2 to 3 for proper functioning.

Bio Bits

A Dr. Søren Sørensen coined the term pH from *pondus hydrogenii,* the Latin for "potential hydrogen." Both letters were originally uppercase, but the style changed to the current pH. Although you hardly ever see reference to the term's origins, it arises directly from a consideration of the hydrogen ion concentration in the solution, or potential hydrogen.

It's Elemental, My Dear

In addition to specific pH requirements, every one of the different living things you see around you—trees, animals, fungi—and those you don't see, such as bacteria, share a few other characteristics in common. One of these characteristics is the SPHONC. Amazing as it may seem, we all consist primarily of **s**ulfur, **p**hosphorus, **h**ydrogen, **o**xygen, **n**itrogen, and **c**arbon. There is a smattering of other elements, such as calcium and potassium, but the SPHONC covers most of it. In fact, across organisms, the proportions of each of these elements are roughly similar.

The Essential Elements

Our best illustrations of the relative abundance of the SPHONC in organisms are the molecules that make up living things. Sugars consist of carbon, hydrogen, and oxygen. DNA consists of carbon, hydrogen, oxygen, phosphorus, and nitrogen. Fats are mostly carbon and hydrogen, and the basic building block of proteins consists of carbon, hydrogen, nitrogen, and oxygen, with an occasional sulfur thrown in.

An almost infinite variety of arrangements of some of these molecules produces the almost infinite variety of life here on Earth. But the atom that is literally and figuratively at the center of them all is carbon.

The Trace Elements

In addition to the SPHONC and a few ions like calcium, sodium, and potassium, we require a handful of other elements to survive. Because these elements occur in us in tiny amounts, we call them trace elements. An example of a trace element necessary to human existence is iodine. Our thyroid requires iodine to process and use thyroid hormone appropriately.

Other necessary trace elements include iron, which we use in transporting oxygen in our blood, and zinc, which plays several roles in immunity and molecular signaling. Metals like copper, cobalt, and chromium also are important in trace amounts.

Carbon: Useful for Many Occasions

If you look at the components of the previous molecules, you'll notice that carbon is their commonality. A peek at their structure also shows that carbon serves as a backbone here, a central molecule there, always in the middle of everything. Every living organism consists of large molecules containing a lot of carbon. Why is carbon so central to life on Earth?

The answer lies in its versatility. Remember those four empty electron spaces in carbon's outer shell? In Chapter 2, we described how carbon could use those spaces to bond with four hydrogens. Each of those bonds was a single covalent bond. But carbon can do much more than that, as the figure shows. And because its electron openings are so evenly spaced around the atom, it can form these bonds with some enormous molecules, resulting in infinite possibilities for creating the building blocks of life.

Carbon can form a variety of different covalent bonds in different combinations.

Carbon Goes with Almost Anything

In chemical color shorthand, carbon is always represented as being black (oxygen is red, hydrogen is blue). Black is an appropriate choice because, as any fashionista will tell you, black goes with almost anything. That's a pretty apt way to describe carbon and its versatility.

A comparison of carbon's bonding possibilities and that of the other elements of life is probably the best way to grasp why versatile carbon is, literally, the central element of life. Nothing else—not sulfur, hydrogen, phosphorus, oxygen, or nitrogen—has the flexibility of carbon. Hydrogen, for example, can form only one covalent bond. Oxygen is limited to two, as is sulfur. Nitrogen and phosphorus can do more but still not as much as carbon.

Even though the other elements can't do what carbon does, they are still an integral part of life. Although there are many ways to bring together the SPHONC elements, they frequently occur in arrangements that give such a specific character to anything that includes them that we call them functional groups.

Special Functional Carbon Accessories

A couple of these functional groups may have names that sound slightly familiar. Amino groups, or $-NH_2$, are a component of amino acids, the building blocks of proteins. Phosphates, consisting of a phosphorus and four oxygens, are a repeating component of nucleic acids, like DNA.

Name	Structure
Hydrogen	$-H$
Hydroxyl	$-O-H$
Carboxyl	
Amino	
Phosphate	
Methyl	
Sulfhydryl	$-S-H$

Biological molecules are built around carbon and use these basic structures for elaboration. Each group confers its special characteristics to whatever molecule it joins. You'll see these again and again in biology.

The addition of a functional group to a molecule gives that molecule a specific character, or function. An amino acid can behave as a base, while a phosphate adds charge to a molecule and transfers chemical energy among molecules. The figure on the previous page gives some of the basic functional groups of life and their structures.

The Least You Need to Know

♦ Water is less dense as a solid than a liquid, making it unique and necessary in the existence of life on Earth.

♦ Thanks to hydrogen bonds, water exhibits adhesion to other water molecules and cohesion to polar surfaces.

♦ The pH scale reflects indirectly the concentration of H+ in a solution, which is acidic at pH <7, basic at pH >7.

♦ Nature uses six atoms to build most biomolecules: sulfur, phosphorus, hydrogen, oxygen, nitrogen, and carbon.

♦ Carbon has special versatility because it can form a variety of covalent bonds in different combinations.

♦ The variety of life arises in part from functional groups added to carbon to build molecules for different purposes.

Chapter **4**

The Big 4: Large Biological Molecules

In This Chapter

- The building blocks of biological molecules
- Carbs for energy, structure, and communication
- Lipids for energy, structure, and communication
- Proteins for many purposes
- DNA and RNA, the nucleic acids

We encountered functional groups and the SPHONC in Chapter 3. The functional groups, assembled into building blocks on backbones of carbon atoms, can be bonded together to yield large molecules that we classify into four basic categories. These four molecules, in many different permutations, are the basis for the diversity that we see among living things.

These Big 4 biological molecules are carbohydrates, lipids, proteins, and nucleic acids. They have many roles, from giving an organism structure to being involved in one of the millions of processes of living. Let's meet each category individually and discover the basic roles of each in the structure and function of life.

Carbohydrates

You've met carbohydrates. We refer to them casually as "sugars," molecules made of carbon, hydrogen, and oxygen. A sugar molecule has a carbon backbone, usually five or six carbons in the ones we discuss here, but it can have as few as three. Sugar molecules can link together in pairs or in chains or branching "trees," either for structure or energy storage.

Bio Basics _____

Some molecular building blocks have several components in common and can form long strings, or backbones, of these common components. For example, every nucleic acid building block has a phosphate and a sugar, regardless of the type of nucleic acid, and you may see frequent reference to the "sugar-phosphate backbone."

On a nutrition label, you'll see reference to "sugars." That term includes carbohydrates that provide energy, which we get from breaking the chemical bonds in the sugar. It also includes sugars that give structure to a plant, which we refer to generally as fiber. Both are important nutrients for people.

Bio Basics _____

Note that all carbohydrate names end in *–ose* (glucose, lactose, fructose, etc.). That's science shorthand for "sugar." If you see a word in biology that ends in *–ose*, go ahead and call it a carbohydrate.

Sugars serve many purposes. They give crunch to the cell walls of a plant or the exoskeleton of a beetle and chemical energy to the marathon runner. When attached to other molecules, like proteins or fats, they aid in communication between cells. But before we get any further into their uses, let's talk structure.

Sugar Structure

The sugars we encounter most in basic biology have their five or six carbons linked together in a ring. Without diving deep into organic chemistry, there are a couple of essential things to know about the standard representations of these molecules.

Check out the sugars depicted in the following figure. The top-left molecule, glucose, has six carbons. (They're numbered, so you can understand the information that comes later.) The sugar to its right is the same glucose, with all but one letter *C* removed (number 6). The other five carbons are still there but are inferred using the conventions of organic chemistry: where there's a corner, there's a carbon unless otherwise indicated. It might be a good exercise for you to add in a *C* over each corner so that you gain a good understanding of this convention. You should end up adding in five carbon symbols; the sixth is already given because that is conventionally included when it occurs outside of the ring.

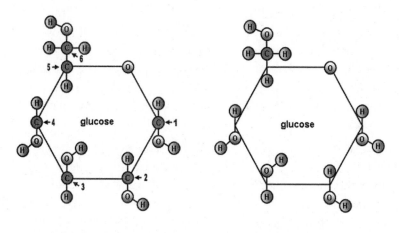

On the left is a glucose, with its carbons indicated. On the right is the same molecule without the carbons indicated (except for the sixth). On the bottom left is ribose, the sugar found in RNA. The sugar on the bottom right is deoxyribose. Note that at carbon 2 (*), ribose and deoxyribose differ by a single oxygen.

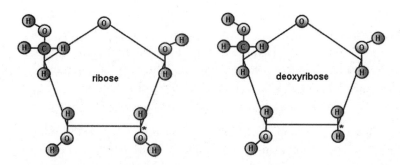

The lower-left sugar in the figure is a ribose. In this depiction, the carbons, except the one outside of the ring, are not indicated or numbered. This is the standard way sugars are presented in texts. How many carbons are in this sugar? Count the corners and don't forget the one that's already indicated!

If you said five, you're right. Ribose is a pentose (*pent* = five) and happens to be the sugar present in *ribo*nucleic acid, or RNA. Guess what the sugar might be in *deoxyribo*nucleic acid, or DNA. If you guessed deoxyribose, you're right.

The fourth sugar in the figure is a deoxyribose. In organic chemistry, each carbon also has a specific number, which becomes important in discussions of nucleic acids.

To count carbons in these sugars, start with the carbon to the right of the non-carbon corner of the molecule. The deoxyribose or ribose always looks to me like a little cupcake with a cherry on top. The "cherry" is an oxygen. To the right of that oxygen, we start counting carbons, so that corner to the right of the "cherry" is the first carbon. Now keep counting. Here's a little test: what's hanging down from carbon 2 of the deoxyribose?

If you said a hydrogen (H), you're right! Now compare the deoxyribose and ribose. Do you see the difference at carbon 2 of each sugar? You'll see that the carbon 2 of ribose has an –OH, rather than an H. Deoxyribose is so named because the O on the second carbon of the ribose has been removed, leaving a "deoxyed" ribose. This tiny distinction between the sugars of DNA and RNA is significant enough that we use it to distinguish the two nucleic acids.

In fact, these subtle differences in sugars mean big differences for many biological molecules. Apparently small changes in a sugar molecule can mean big changes in its function, making the difference between a delicious sugar cookie and the crunchy exoskeleton of a dung beetle.

Sugar and Fuel

A marathon runner keeps fuel on hand in the form of "carbs," or sugars. These fuels provide the marathoner's straining body with the energy it needs to keep the muscles pumping. When we take in sugar like this, it often comes in the form of glucose molecules attached together in a *polymer* called starch. We are especially equipped to start breaking off individual glucose molecules the minute we start chewing on a starch.

Our bodies can then rapidly take the single molecules, or *monomers*, into cells and crack open the chemical bonds to transform the energy for use. The bonds of a sugar are packed with chemical energy that we capture to build a different kind of energy-containing molecule that our muscles access easily. Most species rely on this process of capturing energy from sugars and transforming it for specific purposes.

def•i•ni•tion

A **monomer** is a building block (*mono* = one) and a **polymer** is a chain of monomers. With a few dozen monomers or building blocks, we get millions of different polymers. That may sound nutty until you think of the infinity of values that can be built using only the numbers 0 through 9 as building blocks or the intricate programming that is done using only a binary code of zeros and ones in different combinations.

Polysaccharides: Fuel and Form

Plants use the Sun's energy to make their own glucose, and starch is a plant's way of storing up that sugar. Potatoes, for example, are quite good at packing away tons of glucose molecules and are known to dieticians as a "starchy" vegetable. The glucose molecules in starch are packed fairly closely together. A string of sugar molecules bonded together through *dehydration synthesis*, as they are in starch, is a polymer called a polysaccharide (*poly* = many; *saccharide* = sugar). When the monomers of the polysaccharide are released, as when our bodies break them up, the reaction that releases them is called *hydrolysis*.

Although plants make their own glucose and animals acquire it by eating the plants, animals can also package away the glucose they eat for later use. Animals, including humans, store glucose in a polysaccharide called glycogen, which is more branched than starch. In us, we build this energy reserve primarily in the liver and access it when our glucose levels drop.

def•i•ni•tion

The reaction that hooks one monomer to another in a covalent bond is called **dehydration synthesis**, because in making the bond—synthesizing the larger molecule—a molecule of water is removed (dehydration). The reverse is **hydrolysis** (*hydro* = water; *lysis* = breaking), which breaks the covalent bond by the addition of a molecule of water.

Whether starch or glycogen, the glucose molecules in the polymer are all oriented the same way. If you view the sixth carbon of the glucose as a "carbon flag," you'll see in the figure on the next page that all of the glucose molecules in starch are oriented with their carbon flags on the upper left.

Sugars also serve as structural molecules in a huge variety of organisms, including fungi, bacteria, plants, and insects. The primary structural role of a sugar is as a component of the cell wall, giving the organism support against gravity. In plants, the familiar old glucose molecule serves as one building block of the plant cell wall, but with a catch: the molecules are oriented in an alternating up-down fashion. The resulting structural sugar is called cellulose, a component of fiber.

Bio Basics

Cows live off of plants, but how can they if cellulose is so indigestible for furry critters? The cow's secret is bacteria that have the right proteins to break down the cellulose into digestible bits. This instance is our first of many showing the importance of molecular fit and molecular recognition in biology—and a great example of a mutually beneficial relationship between species.

The orientation of monomers of glucose in polysaccharides can make a big difference in the use of the polymer. The glucoses in the top molecule are oriented "up" and form starch. The glucoses in the bottom molecule alternate orientation to form cellulose, which functions quite differently from starch.

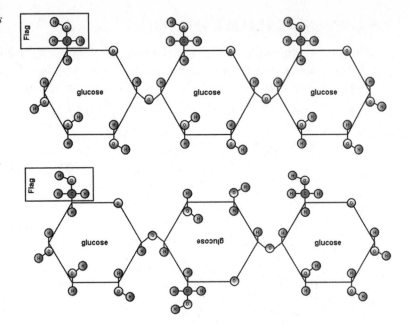

That simple difference in orientation means the difference between a polysaccharide as fuel for us and a polysaccharide as structure. In insect exoskeletons, the building block is also glucose, arranged in an alternating conformation as in cellulose. But in insects, each glucose has an N-acetyl group tacked on, producing a structural molecule that gives bugs that special crunchy sound when you accidentally—ahem—step on them.

These variations on the simple theme of a basic carbon-ring-as-building-block occur again and again in biological systems. In addition to roles in structure and as fuel, sugars also play a role in function. The attachment of subtly different sugar molecules to a protein or a lipid helps cells communicate chemically in a specialized, sugar-based vocabulary. Typically, cells display these sugary messages to the outside world, making them available to other cells that speak the molecular language.

Lipids: The Fatty Trifecta

Starch makes for good, accessible fuel, something that we immediately attack chemically and break up for quick energy. But fats are energy that we bank away for a good long time and break out in times of deprivation. Like sugars, fats serve several purposes, including as a dense source of energy and as a universal structural component of cell membranes everywhere.

Fats: The Good, the Bad, the Neutral

Turn again to a nutrition label, and you see a few references to fats, also known as lipids. (Fats are slightly less confusing than sugars in that they have only two names.) The label may break down fats into categories, including *saturated fats*, *unsaturated fats*, and *cholesterol*. You may have learned that trans fats are "bad" and that there is good cholesterol and bad cholesterol, but what does it all mean?

Let's start with what we mean when we say saturated fat. The question is, saturated with what? There is a specific kind of dietary fat called the triglyceride. As its name implies, it has a structural motif in which something is repeated three times. That something is a chain of carbons and hydrogens, hanging off in triplicate from a head made of glycerol. Those three carbon-hydrogen chains, or *fatty acids*, are the "tri" in a triglyceride. Chains like this can be many carbons long.

Triglycerides come in several forms. You may recall that carbon can form several different kinds of bonds, including single bonds, as with hydrogen, and double bonds, as with itself. A chain of carbons and hydrogens can have every single available carbon bond taken by a hydrogen in single covalent bonds. This scenario of hydrogen saturation yields a saturated fat, saturated to its fullest with every covalent bond taken by hydrogens single bonded to the carbons.

def•i•ni•tion

We call a **fatty acid** a fatty acid because it's got a carboxylic acid attached to a fatty tail. A triglyceride consists of three of these fatty acids attached to a molecule called glycerol. Our dietary fat primarily consists of these triglycerides.

Saturated fats have predictable characteristics. They stick to each other easily, meaning that at room temperature, they form a dense solid. You will realize this if you find a little bit of fat on you to pinch. Does it feel pretty solid? That's because animal fat is saturated fat. The fat on a steak is also solid at room temperature, and in fact, it takes a pretty high heat to loosen it up enough to become liquid. Animals are not the only organisms that produce saturated fat—avocados and coconuts also are known for their saturated fat content.

You can probably now guess what an unsaturated fat is—one that has one or more hydrogens missing. Instead of single bonding with hydrogens at every available space, two or more carbons in an unsaturated fat chain will form a double bond with carbon, leaving no space for a hydrogen. Because some carbons in the chain share two pairs of electrons, they physically draw closer to one another than they do in a single bond. This tighter bonding results in a "kink" in the fatty acid chain.

In a fat with these kinks, the three fatty acids don't lie as densely packed with each other as they do in a saturated fat. The kinks leave spaces between them. Thus, unsaturated fats are less dense than saturated fats and often will be liquid at room temperature. A good example of a liquid unsaturated fat at room temperature is canola oil.

Why do we take in fat, anyway? Fat is a necessary nutrient for everything from our nervous system to our circulatory health. It also, under appropriate conditions, is an excellent way to store up densely packaged energy for the times when stores are running low. We really can't live very well without it.

> **Bio Bits**
>
> A triglyceride can have up to three different fatty acids attached to it. Canola oil, for example, consists primarily of oleic acid, linoleic acid, and linolenic acid, all of which are unsaturated fatty acids with 18 carbons in their chains.

Phospholipids: An Abundant Fat

You may have heard that oil and water don't mix, and indeed, it is something you can observe for yourself. Drop a pat of butter—pure saturated fat—into a bowl of water and watch it just sit there. Even if you try mixing it with a spoon, it will just sit there. Now drop a spoon of salt into the water and stir it a bit. The salt seems to vanish. You've just illustrated the difference between a water-fearing (hydrophobic) and a water-loving (hydrophilic) substance.

Generally speaking, compounds with an unequal electron sharing (ions or anything with a polar covalent bond) are hydrophilic. That charge or an unequal electron sharing gives the molecule polarity that allows it to interact with water through hydrogen bonds. A fat, however, consists largely of hydrogen and carbon in those long chains. Carbon and hydrogen have roughly equivalent electronegativities, and their electron-sharing relationship is relatively nonpolar. Fat, lacking in polarity, doesn't interact with water. As the butter demonstrated, it just sits there.

> **Bio Basics**
>
> Hydrophobic molecules are typically nonpolar (don't form hydrogen bonds) and interact well with each other but not with charged molecules or ions. Hydrophilic molecules are usually polar or ions; they can participate in hydrogen bonding and generally interact well with other polar molecules or ions but not with nonpolar molecules.

One exception to that little maxim about fat and water is the phospholipid. This lipid has a special structure that makes it just right for the job it does: forming the membranes of cells. A phospholipid consists of a polar phosphate head—P and O don't share equally—and a couple of nonpolar hydrocarbon tails. If you look at the following figure, you'll see that one of the two tails has a little kick in it, thanks to a double bond between the two carbons there.

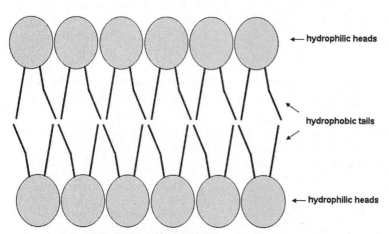

watery environment outside the cell

←— hydrophilic heads

hydrophobic tails

←— hydrophilic heads

watery environment inside the cell

Phospholipids form a double layer and are the major structural components of cell membranes. The molecules are bipolar, with hydrophilic heads and hydrophobic tails.

The kick and the bipolar (hydrophobic and hydrophilic) nature of the phospholipid make it the perfect molecule for building a cell membrane. A cell needs a watery outside to survive. It also needs a watery inside to survive. Thus, it must face the inside and outside worlds with something water friendly. But it also must protect itself by providing a barrier against unwanted intruders.

Phospholipids achieve it all. They assemble into a double layer around a cell but orient to interact with the watery external and internal environments. In the layer facing the inside of the cell, the phospholipids orient their polar, hydrophilic heads to the watery inner environment and their tails away from it. In the layer to the outside of the cell, they do the same.

As the figure shows, the result is a double layer of phospholipids with each layer facing a polar, hydrophilic head to the watery environments. The tails of each layer face one another. They form a hydrophobic, fatty moat around a cell that serves as a general gatekeeper, much in the way that your skin functions. Charged particles cannot simply slip across this fatty moat because they can't interact with it. And to keep the fat fluid, one tail of each phospholipid has that little kick, giving the cell membrane a fluid, liquidy flow and keeping it from being solid and unforgiving at temperatures in which cells thrive.

Steroids: Here to Pump You Up?

Our final molecule in the lipid fatty trifecta is cholesterol. As you may have heard, there are a few different kinds of cholesterol. The "good" cholesterol, high-density lipoprotein, or HDL, in part helps us out because it removes the "bad" cholesterol, low-density lipoprotein, or LDL, from our blood. LDL is associated with inflammation of the lining of the blood vessels, which can lead to a variety of health problems.

But cholesterol has other duties. One of its roles is maintaining cell membrane fluidity. Inserted throughout the lipid bilayer, cholesterol serves as a block to the fatty tails that might otherwise stick together and become a bit too solid.

Cholesterol's other starring role is as the starting molecule for a class of hormones called steroids or steroid *hormones*. With a few snips here and additions there, cholesterol can be changed into the steroid hormones progesterone, testosterone, or estrogen. These molecules look similar but have very different roles in organisms. Testosterone, for example, generally masculinizes vertebrates (animals with backbones), while progesterone and estrogen regulate the ovulatory cycle.

def•i•ni•tion

A **hormone** is a blood-borne signaling molecule. It can be lipid based, like testosterone, or a short chain of amino acids, like insulin.

Proteins

As you progress through this book, one thing will become clear: most cells function primarily as protein factories. It may surprise you to learn that proteins, which we often talk about in terms of food intake, are the fundamental molecule of many of life's processes. Enzymes, for example, form a single broad category of proteins, but there are millions of them, each governing a small step in the molecular pathways of life.

Levels of Structure

Amino acids are the building blocks of proteins. A few amino acids strung together form a peptide, while many many peptides linked together form a polypeptide. When many amino acids in a string interact to form a molecule that's all folded up on itself correctly, we call that molecule a protein.

If we could see a final, properly folded protein with the naked eye, it might look a lot like a wadded-up string of pearls, but that "wadded up" look is misleading. Protein folding is carefully regulated and determined at its core by the amino acids in the chain: their hydrophobicity and hydrophilicity and how they interact together.

For a string of amino acids to ultimately form an active protein, they must first be assembled in the correct order. The simple, unfolded chain is the primary structure of the protein.

 Bio Basics _____

In biology, everything is about molecular recognition through the right fit. To work properly, a protein must be folded up perfectly so that the right parts of it can interact with the right parts of other molecules. For a protein to be wadded up or folded correctly, several factors must come together exactly so, including pH and temperature.

This chain can consist of hundreds of amino acids interacting all along the sequence. Some are hydrophobic and some are hydrophilic. Hydrophobic amino acids interact with one another, and hydrophilic amino acids interact together. Conversely, hydrophobic and hydrophilic repel each other. With these interactions, different configurations arise along different parts of the chain, producing the protein's secondary structure.

Now the protein can fold into its final, or tertiary, structure, ready to actively participate in cellular processes. To achieve tertiary structure, the secondary interactions must usually be ongoing, and the pH, temperature, and salt balance must be just right. This tertiary folding takes place through interactions of the secondary structures along the amino acid chain.

A complete protein can consist of two or more interacting strings of amino acids. A good example is hemoglobin in red blood cells. It grabs oxygen and delivers it to the body's tissues. A complete hemoglobin protein consists of four separate amino acid chains properly folded into their tertiary structures and interacting as a single unit. In cases involving two or more interacting amino acid chains, the final protein has a quaternary structure. Some proteins can consist of as many as a dozen interacting chains, behaving as a single protein unit.

A Plethora of Purposes

What does a protein do? Let us count the ways. Really, that's almost impossible because they do just about everything. Some of them tag things. Some of them destroy things. Some of them protect. Some mark cells as belonging to "self" as opposed to being foreigners. Some serve as structural materials, while others function as highways or motors. They aid in communication, operate as signaling molecules, transfer molecules and cut them up, and interact with each other in complex, inter-related pathways to build things up and break things down. They regulate genes and package DNA, and they regulate and package each other.

As previously described, proteins are the final folded arrangement of a string of amino acids. One way we obtain these building blocks for the millions of proteins our bodies make is through our diet. When we take in those proteins, we can break them apart and use their amino acids to build proteins of our own.

Nucleic Acids

How does a cell know which proteins to make? It has a code, one that is especially guarded in a cellular vault in our cells called the nucleus. This code is deoxyribo-nucleic acid, or DNA. The cell makes a copy of this code and sends it to specialized structures that read it and build proteins based on what they read. As with any code, a typo—a mutation—can result in a message that doesn't make as much sense. When the code gets changed, sometimes, the protein that the cell builds using that code will be changed, too.

Biohazard!

Beware the confusion associated with nucleic acid terms. The shorter term (10 letters, 4 syllables), *nucleotide*, refers to the smaller molecule, the three-part building block. The longer term (12 characters, including the space, and 5 syllables), *nucleic acid*, which is inherent in the names DNA and RNA, designates the big, long molecule.

DNA vs. RNA: A Matter of Structure

DNA and its nucleic acid cousin, ribonucleic acid, or RNA, are both made of the same kinds of building blocks. These building blocks are called nucleotides and consist of three parts: a sugar (ribose for RNA and deoxyribose for DNA), a phosphate, and a nitrogenous base. In DNA, every nucleotide has identical sugars and phosphates, and in RNA, the sugar and phosphate are also the same for every nucleotide.

So what's different? The nitrogenous bases. DNA has a set of four to use as its coding alphabet. These are the purines (adenine and guanine) and the pyrimidines (thymine and cytosine). The nucleotides are abbreviated by their initial letters as A, G, T, and C. From variations in the arrangement and number of these four molecules, all of the diversity of life arises: you, bacteria, wombats, and blue whales.

RNA is also basic at its core, consisting of only four different nucleotides. In fact, it uses three of the same nitrogenous bases as DNA—A, G, and C—but it substitutes a base called uracil (U) where DNA uses thymine. Uracil is a pyrimidine.

Nitrogenous bases pair with each other, using hydrogen bonds, in a predictable way. An adenine will almost always bond with a thymine in DNA or a uracil in RNA, and cytosine and guanine will almost always bond with each other. This pairing allows the cell to use a sequence of DNA and build either a new DNA sequence, using the old one as a template, or an RNA sequence, to make a copy.

DNA vs. RNA: Function Wars

DNA holds the code for proteins. RNA is really a nucleic acid jack-of-all-trades. It not only serves as the copy of the DNA but also operates in reading that copy and building proteins from it. At one point in this process, the three types of RNA come together in protein assembly to make sure the job is done right.

The Least You Need to Know

- The four basic categories of molecules for building life are carbohydrates, lipids, proteins, and nucleic acids.

- Carbohydrates serve many purposes, from energy to structure to chemical communication, as monomers or polymers.

- Lipids, which are hydrophobic, also have different purposes, including energy storage, structure, and signaling.

- Proteins, made of amino acids in up to four structural levels, are involved in just about every process of life.

- The nucleic acids DNA and RNA consist of four nucleotide building blocks, and each has different purposes.

Part **2**

The Cell

The basic unit of life is the cell. As a living protein factory, it needs materials and energy. As a living thing, it must use things, discard things, and recycle things. Cells rely on processes involving gradients, transfers, deliveries, storage, and sugar to get it all done. The original energy source for most of these processes is the Sun. Some organisms can capture this energy to build the molecules that cells need to continue as living things. Often, like any other living things, cells communicate with each other. Also like all living things, they must reproduce. So read on to learn more about the lives of these complex little protein factories.

5

The Cell as Protein Factory: A Guided Tour

In This Chapter

- ◆ The plasma membrane
- ◆ Prokaryotes versus eukaryotes
- ◆ Organelles
- ◆ Movement into and out of the cell

Welcome to the protein factory. You've come for a guided tour of the place where proteins are made, the cell. This factory tour is primarily inside a cell like those that make up your own body, cells with different departments for different processes.

Checking in with Security: The Membrane

Remember the phospholipids from Chapter 4? Their job was to do two things at once: interact with the watery inner and outer worlds of the cell through the hydrophilic heads while protecting the cell from intruders and blocking escapees with the fatty, inner moat. They're obviously important in safeguarding the cell, but as with any good security system, the cell has more than one layer of protection.

In addition to phospholipid protection, the cell has specific guard stations where molecules wanting in or out must obtain permission via chemical recognition to do so. The cell, like any factory, must exchange goods and services and communicate with the outside world.

Biohazard!

The terms for the membrane surrounding the cell can be confusing. Some texts may refer to it as the cell membrane while others may call it the plasma membrane. You also should avoid confusing this membrane with the cellular structure found in plants and some other organisms called the cell wall.

Many of these other security staff members are proteins that span the entire membrane, sometimes with sugars attached to the part that extends to the outside world, as in the figure. I refer to these sugary attachments as "sugar trees" (bio professors everywhere will cringe) because they are sugar polymers that are branched, like trees. The real term for the protein + sugar tree combination is glycoprotein (*glyco* = sugar). Yes, the cell uses sugar trees to talk with other cells or messenger molecules.

Proteins can span the plasma membrane and communicate with the inside and outside environments of the cell. Often, specific branched sugar polymers ("sugar trees") attached to the protein extend from the cell to communicate chemically with other cells or messenger molecules.

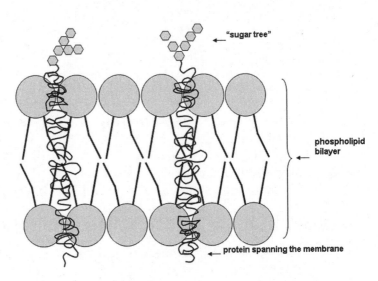

watery environment outside the cell

"sugar tree"

phospholipid bilayer

protein spanning the membrane

watery environment inside the cell

The plasma membrane consists of the phospholipid bilayer, embedded proteins that may span the entire bilayer, and cholesterol in animal cells. Thanks to those "kicks" in the phospholipids (see Chapter 4) and to the cholesterol, this collation of molecules is not an impenetrable solid but instead is what scientists call a fluid mosaic. If you melted a peanut/chocolate/caramel candy bar and let it cool just a bit, the results would be a good representation of the fluid mosaic—the slowly flowing chocolate with a mosaic of proteins, fats, and sugars densely distributed in it.

Some organisms—including plants, some bacteria and Archaea, and fungi—have a cell wall. The cell wall is not the same thing as the plasma membrane and is a separate structure consisting of sugars and other molecules. We've encountered a main component of plant cell walls previously in the form of cellulose (see Chapter 4).

Bio Bits

The plasma membrane is a busy place. In addition to its proteins, sugars, cholesterol, and gases and small molecules moving in and out, you'll also find phospholipids spinning, moving from side to side, and sometimes even flipping over. Rather than being a sort of static gelatin mould with molecules immobilized in it, the plasma membrane is a fluid mosaic of movement and constantly shifting components, like a bustling entry to a metropolitan subway station.

Nucleus Optional: Prokaryotes and Eukaryotes

The life family tree contains two basic kinds of cells, but variations on these two themes result in the millions of species the earth supports. The first, and older, kind of cell is called a *prokaryote*. *Pro* means before, and *kary* means kernel or nucleus. Thus, you can think of these cells as being around before anything had a nucleus. These nucleus-free cells are the most commonly occurring cell type on Earth and actually make up two of the three Domains of life, Bacteria and the Archaea, the weirdos of the prokaryotic world.

Bio Bits

Abundant research in recent decades has revealed enormous differences between the outwardly similar single-celled Bacteria and Archaea prokaryotes. We now know that Archaea are genetically as different from Bacteria as Bacteria are from us and other eukaryotes.

The second basic cell type, the eukaryotic cell, characterizes the third Domain, Eukarya. *Eu* is the Greek for true, so this term means true nucleus. True to that name, all organisms consisting of eukaryotic cells—called eukaryotes—generally have nucleated cells. Eukaryotes include humans, geese, hammerhead sharks, yeast, and single-celled amoebas. In fact, all animals, fungi, plants, and protists are eukaryotes.

These two cell types have some things in common. (They have to, as cells and protein factories.) They also, however, have many differences beyond the key feature of the presence or absence of a nucleus.

How They Are Similar

Every cell has a plasma membrane. They all have DNA and make proteins. They have *ribosomes* and, in the case of bacteria, plants, fungi, and some Archaea, cell walls. Their insides contain the jelly-like, mostly water substance called cytosol, a component of the cytoplasm. They also both have proteins that form an infrastructure called a cytoskeleton, although each cell type uses cytoskeleton proteins for different purposes.

def•i•ni•tion

A **ribosome** is a structure in the cytoplasm, consisting of RNA and associated proteins. It "reads" the instructions for building proteins.

That's it. Unless you count the enormous diversity of organisms represented by each cell type—and that's an important similarity, too—that's the entire list of basic commonalities between prokaryotes and eukaryotes.

Biohazard!

Don't fall into the trap of using the terms cytoplasm and cytosol interchangeably. They are technically not the same thing. The cytosol is the liquid component of the inner cell outside of membrane-bound organelles. It is, in fact, a component of the cytoplasm. The cytoplasm more generally refers to anything lying within the cell membrane, including membrane-bound organelles.

How They Differ: A Longer Story

Their differences make a longer list. For starters, eukaryotes are much larger, as much as 100 times bigger than a prokaryote. If you Google for images of "macrophages," for example, you'll see the big difference: often these images show a macrophage—a large eukaryotic immune cell—extending part of its cell membrane to capture a much smaller rod-shaped bacterium.

The size difference is related to two things. For the first, imagine a cell that requires nutrients at its center, as cells do. Prokaryotes have no good transport systems to get nutrients over big distances, so they must be very small for the needed chemicals to move from the cell periphery to the center. They have a low surface-to-volume ratio as a result.

> **Bio Bits**
>
> Not all cells in us eukaryotes keep the nuclei they're born with. As with most mammals, our red blood cells, which carry oxygen, lack a nucleus when mature. The missing nucleus makes for more efficient oxygen transport. In fact, the presence of nucleated blood cells in a human indicates disease.

Bio Basics

As the security guard monitoring nutrient movement, managing waste, and transmitting messages, the plasma membrane serves the entire cell, all the way to the center. Bacteria, lacking a rapid-delivery system, must remain small so molecules can reach the center in good time. But eukaryotic cells have compartmentalized processes and a highway system for rapid transport, allowing them a large volume relative to the surface available to manage all that traffic.

The second reason is related to the first. Given our model of the cell as a protein factory, eukaryotes and prokaryotes are very different kinds of factories. The prokaryote is the equivalent of a small mom-and-pop operation, with all the plans, parts, and workers gathered around a central area. But a eukaryotic cell is like a huge, bustling production plant with different areas cordoned off for different functions: the generator here, the waste plant over there, the assembly line just outside the nucleus. And it has cellular superhighways for transporting everything across its relatively huge distances.

Inside the eukaryotic cell's nucleus is the DNA, which I just mentioned as a commonality between prokaryotes and eukaryotes. However, although they both use DNA as genetic material, their DNA is different. Prokaryotic DNA is circular, while eukaryotic DNA has a beginning and an end, like a piece of floss. In addition, eukaryotic DNA is packaged in a complex hierarchy of folding around proteins, while most prokaryotic DNA is not associated with this kind of folding hierarchy.

Sorry to muddle things up more, but remember how you just learned that both cell types have ribosomes? It's true, but … their ribosomes differ, too. Eukaryotic ribosomes are far more complex: they have 5 kinds of RNA and about 80 different kinds of proteins, while prokaryotic ribosomes consist of only 3 kinds of RNA and about 50 kinds of proteins.

The final difference relates back to the two cell types as different-size factories. Because it breaks up factory functions into different areas, the eukaryotic cell has specific, membrane-bound structures responsible for these separate functions. The mom-and-pop prokaryotes, with all the work done in the middle of the factory floor, do not have these structures, the organelles.

Membranes Required: Organelles

You'll notice that I referred to these structures as being "membrane bound." Is that membrane the same kind of membrane that we find around the cell itself? Glad you asked. The answer is, "Yep."

Sometimes, people refer to ribosomes as organelles, but there are arguments about the accuracy of this designation. While prokaryotes and eukaryotes do both have ribosomes, which lack membranes, only eukaryotes have membrane-bound organelles, including the one that gives them their name: the nucleus.

Nucleus: The Safe

That kernel, center, karyo of the eukaryote is the factory "safe" where the precious original copy of the code for building proteins is stored. DNA lies carefully packaged and folded up in the membrane-bound nucleus, accessible only to a very select group of molecules. In fact, the nuclear membrane is much like the plasma membrane, except even more picky. It contains nuclear pores, special gates for molecules to use when entering or exiting, but nothing gets through these gates either way unless it has a specific identifying tag. Not even the copy of the code—intended for use outside the nucleus on the factory floor assembly line—can exit without the right ID.

The DNA is so closely protected that it never really leaves the nucleus, although if the cell divides, the nucleus itself briefly disassembles but then reforms as rapidly as possible. And in the nucleus, the DNA isn't unwound from its proteins unless it's absolutely necessary—say, when it needs to be copied into the message that the factory workers will read for building proteins.

Inside the nucleus is also a structure called the nucleolus. While this word signifies "little nucleus," it is not actually a separate, smaller nucleus. It is an area of the nucleus where the cell is busy putting together some of the components of ribosomes.

Mitochondria: The Generator

The mighty mitochondrion (that's the singular form of mitochondria) is an interesting little organelle. It's often called the "powerhouse of the cell" because … well, its job is to provide the power the cell needs for its many processes.

Remember that only eukaryotes have mitochondria. In these kidney-bean-shaped organelles, energy harvested from the glucose you eat is repackaged into energy-packed molecules to fuel the cell's—and your body's—processes. Tissues of your body with high energy requirements also have a lot of mitochondria, including your liver, which is a metabolically busy organ.

Bio Bits

Mitochondria behave as if they once were independent organisms, which may have been the case. They have their own DNA and ribosomes and make many of their own proteins, in addition to having a membrane bilayer. When the cell divides, they also divide by a mechanism similar to that of bacteria. These features have led to the hypothesis that mitochondria may once have lived independently but established a working partnership with larger cells and over time became organelles.

This organelle actually has a two-layer membrane, an inner and an outer membrane, separated from one another by a space, naturally called the intermembrane space. (Biologists may not be that imaginative sometimes, but at least they're concise.) The inner membrane is the site of much activity related to building ATP, often referred to as the "energy currency of the cell." ATP, or adenosine triphosphate, is actually a nucleotide—the building block of a nucleic acid—which you may recall consists of a sugar, a phosphate, and a base. But ATP is a special nucleotide with loads of energy packed into those two extra phosphates added onto it, giving it the "tri" in its name.

Inside the mitochondrion, encompassed by the inner membrane, is the matrix. This doesn't have anything to do with choosing a red pill or a blue pill or alternate realities, but it is the place where the mitochondrion houses its very own DNA and ribosomes. Yes, it's true: This organelle acts a lot like the prokaryotic cell that it likely once was, and it is responsible for making the proteins it uses. When the cell divides, the mitochondria divide, too, doubling their numbers so that there are enough to allocate to each of the two new cells that are formed.

Lysosome: The Garbage Disposal

Not unlike our own digestive system, the cell has an acidic area where "foods" are broken down and waste processed, much of it for recycling. In us, this system includes the acid environment of our stomachs. In the eukaryotic cell, it consists of a membrane-bound organelle called the lysosome. Its name says it all: *lyse* means to break things up, and *soma* means body. Thus, the lysosome is the body that breaks stuff.

This organelle is something the cell produces itself by packing up a nice phospholipid membrane-bound sphere full of special proteins that break up other molecules. An important player in this process is pH (see Chapter 3). Most of the proteins in the cell operate best at a fairly neutral pH, between 7 and 8. But the proteins in a lysosome, called hydrolases, do quite well at a very low pH, around 4.8. As their name implies, they break up molecules through hydrolysis.

This setup serves a safety purpose. Remember that one of the things that ensures proper protein folding is the proper pH. A hydrolase at pH 7 isn't folded right and likely won't break up any proteins. A protein appropriately packaged into a lysosome, however, is at the correctly acidic pH and ready to break things. Thus, if a lysosome accidentally releases hydrolases into the cell, they won't run amok and destroy everything because the higher pH will immediately cause them to unfold. Cell functions are rife with these kinds of well-regulated safety mechanisms.

The lysosome follows the recycle-and-reuse philosophy, breaking big molecules down into their component parts, which then often return to assembly areas in the cell to be used again. Thus, the separate amino acids that result from the disassembly of a protein in the lysosome can be used again to build more proteins.

Chloroplasts: Why Plants Rule the Earth

Not all eukaryotic cells are created equal, and plants and other organisms that photosynthesize have an edge over the rest of us. Photosynthesis involves capturing the Sun's light energy and using it to build sugars out of carbon dioxide and water molecules. This process of capturing carbon dioxide gas and rejiggering the carbon with water molecules to make food takes place in a membrane-bound organelle called the chloroplast.

When a substance reflects specific wavelengths of light, we see that reflected light as a specific color. If you see something red, it's because that item reflects the red wavelengths of light. Because the main pigments in chloroplasts reflect sunlight in the "green" wavelength, these organelles appear green to us. This color is what gives them their name: *chloro* is Greek for green, and *plast* means form.

These "green forms" may, like mitochondria, once have been free-living prokaryotes. Also like the mitochondria, they have an inner and outer membrane with an inter-membrane space lying between the two.

The inside part of the chloroplast is called the stroma, and in the stroma are stacks of flat little pancakelike structures called thylakoids. Each stack of thylakoids is called a granum (plural: grana). In each little thylakoid pancake, photosynthesis takes place. Life on Earth exists in its current form thanks to those photosynthesizing pancakes.

Bio Basics

A common question students have because many texts don't specifically address it is, "Do plant cells have mitochondria?" The answer is, Yes, they do, and they use them for the same purposes as other organisms.

I called this section "Why Plants Rule the Earth" because almost every organism on Earth is tied either directly or indirectly to photosynthesis and thus to the Sun. Plants make their own sugars; organisms eat plants for this sugar, other organisms eat the organisms that eat the plants; many bacteria exist in symbiosis with the plants or rely on their sugars; fungi rely on the dead things that rely on the plants, and on and on. Earth's great chain of being starts with the Sun, which works through the plants and billions of photosynthesizing bacteria. I always like to imagine how things would be if we humans could photosynthesize. It would make mealtimes pretty boring but imagine the savings in conservation terms!

Factory Workflow: An Endomembrane System

The time has come to talk factory workflow. If people modeled their factories on the cellular model, mistakes would be few and far between, thanks to the incessant checks that occur at every step of the process. The details about those regulatory processes could—and have—literally filled a book, so we won't get that deep here. But you do need to understand the flow of the factory areas that are centrally involved in making proteins, the endomembrane system.

The endomembrane system.

It all starts in the nucleus, where the DNA is housed. The cell copies this code for building proteins into RNA and sends that copy out to the factory floor in the cytoplasm. The endoplasmic reticulum (ER) is the assembly line where the ribosomes work to read the code and build proteins (1 in the figure). The ER is really an extension of folded membranes associated with the nuclear membrane.

There are two basic types of ER. The first is studded with ribosomes, each busily reading a code and using it to construct the right protein. Because a ribosome-studded membrane network looks bumpy or rough on microscope images, this ER is called the rough endoplasmic reticulum, or rER. Where rER is, proteins are being made.

The other type of ER is the ribosome-free version, called smooth ER. The sER is the site of many activities, including production of lipids and detoxification of drugs. A special kind of sER, called sarcoplasmic reticulum, occurs in muscle cells. Its job is primarily to manage calcium ion movement when the muscle cells receive nervous system messages about contracting.

From the rER, the protein is packaged into vesicles (2 in the figure), little membrane-bound bags of material. The final stop in the endomembrane system is the Golgi apparatus (also referred to as the Golgi body or simply the Golgi). In the Golgi apparatus (3 in the figure), mysterious and wonderful things happen to proteins, including the addition of the "sugar trees" that cells use in signaling. The Golgi packages the modified proteins into little blebs of membrane called vesicles (4 in the figure). These pinch off of the Golgi with their precious cargo inside and head for their specific destinations.

Vesicles: Bags, Suitcases, Backpacks

All organelles require proteins, some of which they may produce themselves but others that must be sent to them in vesicles. These suitcases of proteins or fats or other molecules travel along the cell's highway system to get to their destinations.

In the eukaryote, the highway system consists of special proteins called actin and tubulin. These proteins form the cytoskeleton, which gives the cell structure and serves as superhighways for transport. The highways form microfilaments (actin), intermediate filaments (other structural proteins), and microtubules (tubulin). Motor proteins attached to the vesicles "walk" or "crawl" along these highways, bound for organelles, other areas of the cytoplasm, or the cell membrane. At the membrane, a molecule's fate can vary from being associated with the inner surface of the membrane to being embedded in the membrane to being ejected into the world outside the cell.

One organelle, the lysosome, begins life as a vesicle packed with hydrolases. After a maturation process, the vesicle officially becomes an Important Cellular Organelle. The lysosome has a feature of vesicles that is important to the jobs that they do: anything made of membrane can fuse with other membranes, much as you could squish two pats of butter together. Thus, a vesicle carrying molecules destined for the outside world can fuse with the cell membrane and release its contents. Or, a lysosome can fuse with a vesicle containing an ingested food particle, exposing the "food" to its enzymes for breakdown.

A vesicle that serves a specific purpose in the cell is called a vacuole. Food vacuoles result from ingestion of a food particle, while plant cells contain a large central vacuole that stores up water for these organisms that cannot go seek it for themselves.

Crossing the Cell Membrane

We started this chapter with the cell membrane, and we're going to wrap up with it, too. That fluid mosaic of proteins, fats, and the occasional sugar tree is often a destination for proteins packaged into vesicles. The proteins embedded in the membrane itself play a major role in security and trafficking into and out of the cell.

Passive Movement: Diffusion

Some molecules enter the cell without the help of proteins. These include small molecules such as gases—oxygen, carbon dioxide—and water. Yes, water, in spite of the repellent fatty moat, can wander freely across the cell membrane. But it will do so down a gradient: if there is an imbalance of water molecules inside compared to outside, water will move from high to low to restore the balance. This movement of molecules from areas of high concentration to low concentration, or *diffusion*, is passive because it requires no energy input. It just happens spontaneously.

def•i•ni•tion

Diffusion is the passive movement of molecules from an area of high concentration to an area of low concentration. The special term for this behavior in water molecules is called **osmosis**.

The diffusion of water across the semipermeable membrane has a special term, *osmosis*. Osmosis refers specifically and only to the movement of water molecules from high to low concentration.

Facilitated Diffusion: I.D. Required

Molecules that are larger or that carry a charge can still get across the membrane passively (provided the gradient requires it). But they need special protein channels that allow them to pass through. Because the proteins that permit this passive movement form channels through the membrane, they are called channel proteins or, more generally, transport proteins. Like I said, biologists may not be inventive about naming things, but at least the names make sense.

Proteins like this span the entire membrane so that the entry and exit points of the channel extend beyond the membrane. For the protein to do this, the parts that fit comfortably through the middle of the phospholipid bilayer are hydrophobic, while the parts that stick out into the watery environments are hydrophilic. Inside the channel, the protein may have specific charges that either attract or repel specific ions.

Because the protein facilitates passive movement of molecules, this kind of movement—which still requires no energy input and relies simply on unequal concentrations—is called facilitated diffusion. Another kind of protein, a carrier protein, facilitates diffusion of larger molecules through chemical recognition that opens the channel to let the molecules in or out.

Active Transport: Energy Investment

A factory as highly regulated as the cell cannot rely solely on concentration gradients to move molecules. Sometimes, the cell must invest energy into the process and push molecules out against their concentration gradient, from low to high concentration.

The same principle operates when we use an air conditioner. If it's cool outside and hot inside, we just open the windows and let the heat dissipate outward passively. But if it's hot outside and we want it to be cool inside, we have to input energy—electrical energy—into turning on the air conditioner, which does the work of cooling the air. Because pushing something against the gradient, from low to high, takes work (energy input), it is referred to as an active process.

In cells, these processes are called active transport. The cell uses up energy-containing molecules like ATP to transport other molecules against their gradient. Usually, this kind of transport is done via proteins in the membrane. Because biologists think of these proteins as pumping molecules out of the cell, they are often referred to as protein pumps.

Bulk Delivery: Exocytosis and Endocytosis

Sometimes the cell needs to take in a bulk delivery of molecules. If these molecules are water or something dissolved in the water, the cell ingests them in a nonspecific process called pinocytosis, which means "cell drinking."

Another nonspecific form of ingestion is phagocytosis, or "big eating." The cell will snag large bits of food or bacteria with extensions of the plasma membrane and fold the food particle or hapless prokaryote into the cell inside a vesicle. Both pinocytosis and phagocytosis are active (energy-requiring) processes called more generally endocytosis.

The reverse of endocytosis is exocytosis. The cell can take molecules packed up in vesicles and send them to the membrane, where they fuse with it. The vesicle, once fused to the membrane, releases its contents into the world outside.

The Least You Need to Know

- ◆ The plasma membrane acts as the first general security checkpoint for the cell.

- ◆ Cells are either prokaryotes, which lack a nucleus and include Bacteria and Archaea, or eukaryotes, with a nucleus.

- ◆ Eukaryotes are large, compartmentalized cells with membrane-bound organelles and a cytoskeleton "superhighway."

- ◆ Protein packaging occurs via protein transport in vesicles from the endoplasmic reticulum to the Golgi and then to their destinations.

- ◆ The cell moves molecules through passive diffusion or active endocytosis and exocytosis.

Chapter 6

Metabolism: Nature Abhors a Gradient

In This Chapter

- ◆ Laws of thermodynamics
- ◆ Exergonic, endergonic, and coupled reactions
- ◆ ATP as the energy currency of life
- ◆ Enzyme recognition and regulation

Any book about biology requires a chapter on energy because every process of life involves some transfer or transformation of energy, from the small scale of molecule building to the large scale of ecosystem dynamics. Just reading that sentence required many transfers of energy in you. But let's not get ahead of ourselves. Here, we start small, with just the basics.

Much of the energy transfer in biological processes occurs because molecules are moving down their concentration gradient, from high to low. Part of this chapter's title is "Nature Abhors a Gradient." I say that because everywhere we look, we find a constant search for equilibrium as molecules rush down their gradient. But the fact is, nature doesn't simply seek to rectify the inequalities of a gradient. Nature also uses these inequalities to drive the processes of living.

Energy In or Out?

The processes of life, collectively known as *metabolism*, are all about paired energy reactions. For an active (energy-requiring) process to proceed, it must acquire energy derived from some other process. Energy released by one process is used to drive another. These reactions are thus what biologists refer to as coupled reactions: energy released is input into something else.

def•i•ni•tion

Metabolism encompasses the chemical processes in an organism that support life and involves the transfer and transformation of energy. **Exergonic** processes involve transfer of energy outward from the reaction.

Typically, an "energy-out" or *exergonic* process involves passive or spontaneous movement from high to low. In the cell, this movement refers to the movement of molecules. In Chapter 5, we talked about how molecules will move from a higher concentration to a lower concentration because of nature's abhorrence of a gradient. Whenever the concentrations are out of balance, look for nature to knock on any door—or try any available membrane protein—as a conduit to correct it and bring the gradient to equilibrium (equal concentrations on both sides of a membrane).

A Law of Thermodynamics

We can't really talk about all this transfer of energy without addressing two important laws of *thermodynamics*. The first is the law that energy cannot be created or destroyed, it can only be transferred or transformed. Thus, we never speak of "creating" energy.

def•i•ni•tion

Thermodynamics, which literally means "movement of heat," is the study of the transfer and transformation of energy, including heat. **Heat** is the workless (passive) transfer of energy from an area of high temperature to an area of low temperature.

Where energy came from in the first place is a matter for the physicists to work out.

But we can, even among biologists, always speak of energy transferring from one thing to another, as *heat* does spontaneously from hot to cold in a workless transfer of energy. And we can speak of energy being transformed, as happens when the positional or potential energy of a rock balanced on the cliff's edge transforms into kinetic energy or the energy of motion when the rock tumbles off of the cliff to the ground.

When that rock falls off of the cliff and its energy transforms from potential to kinetic, something else happens, too. The transformation is not 100 percent efficient. Some of the energy may be transferred as heat because of friction between the rock and the hillside, for example. This loss of some of that useful potential energy as heat is a great example of the second law of thermodynamics.

The second law addresses disorder. Our measure of disorder is entropy, and the law states that with transformation or transfer of energy, the entropy of a system will increase. You'll realize this if you look around your bedroom. You may clean your room, but you know it won't stay that way. As time passes and energy is transferred and transformed through a million little processes of living, the disorder of your room will undoubtedly increase. Entropy has a way of doing that, and after all, it *is* a law. The next time someone requests that you clean up, try using the second law of thermodynamics to explain your way out of it.

Bio Bits

Because we intuitively expect disorder to increase over time, we sometimes refer to the second law of thermodynamics as "Time's Arrow." When you get all bright and shiny to go out at night, you don't expect to return several hours later looking so perfect. A look in the mirror when you return from a night on the town gives the evidence that time has indeed passed and your disorder has increased.

Throughout nature, there is that inefficiency to transfers and transformations of energy. The so-called food chain is a good example. When organisms take in the chemical energy of plant matter and transform it for their own processes, the outcome may be that as little as 10 percent of the energy packed into those plants is actually recouped for the organism that ate the plant. The remaining 90 percent is lost as heat as the energy is transformed.

Energy In: Endergonic Reactions

Pretty much any process that involves building up something involves a net input of energy. Such processes are called endergonic processes (*ender* = into). Combining two amino acids together requires an input of energy to physically bring them near each other and link them. The energy available for use in such a reaction is known as *free energy*.

def•i•ni•tion

The energy from an exergonic reaction that can do work in another reaction is called **free energy**. The specific symbol to indicate free energy is G, named after J. Williams Gibbs, the man who defined it. When the change in G, Δ, is negative, the reaction is exergonic—it releases energy, much of which can be captured to do work.

Endergonic reactions are usually "build 'em up" reactions, which we refer to in fancy terms as anabolic. You may have heard of anabolic steroids—baseball, anyone?—and they're called that because they build up muscle for the people who take them. They also have some pretty undesirable side effects, so don't try that at home.

An easy example of an endergonic reaction in the visible world is building a tower of blocks. You use energy, doing work on the blocks to stack them. The result is a relatively unstable structure with a lot of potential energy packed into it.

An easy example of an endergonic reaction at the molecular level is the summing of two molecular blocks. As the figure on the next page shows, adding a phosphate to ADP (adenosine diphosphate) to make ATP (adenosine triphosphate) is an endergonic, or energy-requiring process: ADP + P = ATP.

Energy Out: Exergonic Reactions

The energy input into an endergonic reaction has to come from somewhere, and that somewhere is usually a nearby exergonic, or energy-out, reaction. These reactions are "break 'em down" processes, known in fancy biological circles as catabolic reactions.

Want to imagine a catabolic reaction that releases energy? Think of that tower of blocks. It's unstable but packed with the potential energy of each block at its height, waiting to fall. When the tower falls, the blocks release that energy. If there is another tower nearby, the released energy can transfer to that tower and knock it down, too. With its energy output, the first tower provided the energy input to knock over the second tower.

For an example in biology, take our ADP + P = ATP buildup in reverse: ATP (and some water) = ADP + P. Why the water? The P is removed from ATP via that old standby breakdown process, hydrolysis. When water breaks the bond holding the third phosphate onto ATP, the chemical (potential) energy packed into the bond can be transferred to a process requiring energy input.

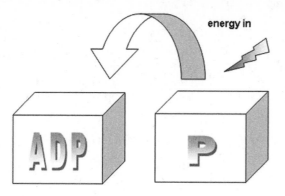

energy in

Bringing ADP and P together requires work, just as stacking real blocks requires doing work on the blocks. Because energy is input, this buildup to ATP is endergonic.

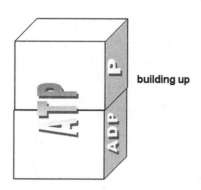

building up

At the completion of an exergonic reaction, the outcome is molecules that contain less potential energy but that are far more stable. Imagine the tower of blocks again, now knocked to the ground. The blocks no longer have that much potential energy, but they are much more stable on the ground than they were in their tower formation. The free energy of this particular group of blocks has decreased considerably as they gained in stability. In the same way, if you climbed on top of a table and stood on the edge, you'd be more unstable but have greater potential energy than if you were standing on solid ground.

Bio Basics

In biological processes, potential energy often exists as the energy packed into unstable chemical bonds, such as the one holding the third phosphate onto ATP. This phosphate is perched unstably at the tip of the molecule, like a rock on the edge of a cliff. Just as tipping the rock over the cliff releases its potential energy, snapping open the high-energy bond holding on the third phosphate of ATP releases that potential energy.

The same applies to ATP, with its unstable third phosphate block added. When that ATP is broken into ADP and P, these two molecules don't contain as much potential energy, but they are far more stable. Their free energy also has decreased, presumably transferred now for use in some other endergonic reaction.

Nature is full of examples of spontaneous reactions like this in which the free energy of the system decreases (because energy is released), and the system becomes more stable but can do less work than before. ATP is unstable but can do a lot of work. ADP and phosphate, having released energy (its system energy thus decreased), are more stable but can do much less work.

Energy Out, Energy In: Coupled Reactions

You eat glucose in your food, which your body transfers into the cell and breaks open for its chemical energy (energy out). This chemical energy is then used to provide the necessary energy to build ATP (energy in), the "energy currency" that your muscles can use when you move. When you move, your body snaps the third phosphate off of ATP, releasing that chemical energy (energy out), which your muscles use in the processes of contraction (energy in).

With every single one of these transfers and transformations of energy, the change-over is inefficient, and much of it is lost as heat. For this reason, when you flail around your arms and legs doing aerobics, you start to feel heated.

These dual reactions involving energy-in/energy-out are the coupled reactions of biology. You encounter this coupling of processes—one releasing energy and one requiring it—over and over again in our study of life.

The Energy Currency of Life: ATP

You've already met ATP as the "energy currency" of the cell (see Chapter 5). Its third phosphate bond holds much of the energy that drives life's processes.

The lowly phosphate is really one of the primary movers and shakers of cellular activity. In Chapter 9, we see that the addition or removal of a phosphate regulates many steps of cell communication. It may seem odd that a little molecule like phosphate—with only its phosphorus atom and four oxygens—plays such a ubiquitous role. But its importance may now make clear why rocks—which are full of phosphates—are perhaps more significant to your existence than you realized.

Many proteins exist solely to break off the third phosphate from ATP. Once that bond breaks, the potential energy packed into it—in the form of chemical energy—is released and can be coupled to some endergonic process that must take place.

What's left once the third phosphate is snapped off? Well, now the molecule is an ADP. As we've mentioned, the cell is a recycle/reuse kind of entity, so it simply takes that depleted ADP and uses a protein specialized for such things to snap on another P and make more ATP. Imagine the stores of phosphate a cell must have ever at the ready to replenish all the depleted energy-currency molecules lying around.

Biohazard!

ATP is not the only energy-currency nucleotide. GTP, or guanosine triphosphate, has a similar role in several well-characterized signaling reactions in the cell. Why is ATP the preferential energy-currency molecule? Some researchers speculate that adenine may have been the first nucleotide that nature synthesized, making ATP the first-place energy candidate by default.

Enzymes and Energy

Imagine two amorphous blobs sitting a foot apart. If they come together, they will interact and become a blob that has some morph (form) to it. But they have no way of coming together to interact. A certain amount of work or input of heat energy is required to get these two together. Unless something helpful comes along, they'd sit there for millennia, not moving and never getting close—that would be a molecular tragedy.

But if you came along and nudged them toward one another, close enough to touch, they'd be able to complete their reaction and play their important roles as blobs. You'd be acting just like an enzyme.

Whenever I ask my students what an enzyme is, many of them can readily respond, "A catalyst." But then we usually get stuck. What is a catalyst? What does a catalyst do? A good place to start is with the original Greek meaning, which was "able to dissolve." But dissolve what?

Enzymes Lower the Energy Bar

The answer is, the energy barrier or activation energy. A specific amount of energy is required for those blobs to get together and react. If you come along and physically bring them closer, that lowers the amount of energy required for them to react, and their reaction is now much more likely to happen and to happen a lot sooner. The enzyme's job is to lower this activation energy, or the energy that must be input into a reaction for it to move forward.

There are two ways for molecules to move together. They can be energized by the transfer of heat, which might increase their kinetic energy (energy of movement) enough for them to run into each other and react, the way two hyper toddlers can bounce off of one another. But often, the temperatures of life are not sufficient for the amount of heat required to achieve this level of motion. That's where *work* comes in. If heat can't do the job, then there must be transfer of free energy to do the work instead.

def•i•ni•tion

In physics (and thus, ultimately, biology), **work** is the transfer of energy by a force acting over a distance. If that makes you think, "What?!", here's another way to think about it. If you push on your desk (applying force) and your desk moves, you've done work on the desk. Enzymes do work on molecules to lower the energy required for them to react.

Either way, if the job is to break bonds, the energy must be sufficient to stress the bonds enough to snap them. If the job is to bring reactions together, the energy must be sufficient to bring the reactants together.

The solid line in the figure on the next page shows the energy barrier or hump that must be overcome before a reaction can take place. The dotted line shows what can happen when an enzyme is present. The enzyme's action lowers the energy barrier to a manageable level, and the reaction occurs. As you can see, this downhill reaction results in the breakdown of the reactants into products with a low free energy, so it is exergonic and free energy is released. Whether the enzyme is present or not, if the reaction is completed, the same amount of free energy will be released by the breakdown of these reactants—they will "fall" the same distance when the reaction occurs.

Bio Basics

Almost all enzymes are proteins. The major exception is when RNA behaves like an enzyme by catalyzing its own reactions. For protein enzymes, the convention is to name them using the suffix *-ase*, as with hydrolases. Whenever you see a term ending in *ase* in biology, think enzyme.

Thanks to enzymes, the world hums along at a decently fast rate of reaction, whether it's adding P to ADP or breaking up a glucose molecule in the cytoplasm. Without enzymes, we'd still be stuck back at "Start" in the game of biological processes.

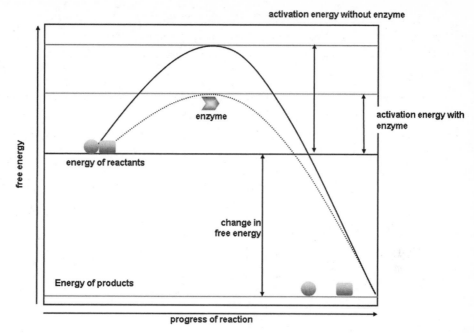

Lowering the energy barrier.

Enzyme Recognition

As you may have come to expect by now, not just any enzyme can do this for just any molecule. For each reactant or reactants, there is a specific enzyme that is folded just right, making a little pocket that allows it to recognize its target chemically. If the key (reactant) doesn't fit the lock (enzyme), then the enzyme can't work with that molecule. The molecule that the enzyme targets with its lock is called its substrate.

The lock where the key, or substrate, is supposed to fit perfectly is called the enzyme's active site. As the figure on the next page shows, for the active site to be active, the enzyme (a protein, remember) must be perfectly folded to fit its target molecule. That means that enzymes, like other proteins, need the right pH, salt, and temperature for best function.

Enzymes must be a fit for their target molecule.

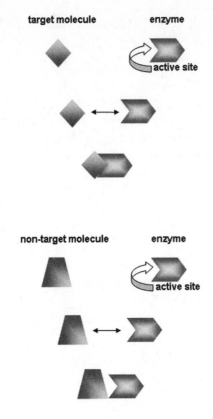

This kind of regulation is excellent when it prevents an enzyme from targeting the wrong reactant. But it also can mean disease when the enzyme is right about its target but can't fit with it. If there is an error in the code (the DNA) for building the enzyme—and errors do happen—then there may be an error in how the enzyme folds or where it's supposed to recognize its target. The faulty protein can't pick up on the reactant and catalyze the reaction—and the reaction may not occur at all.

Unless there is a backup system.

Enzyme Regulation

Just as we're getting used to the fact that the cell regulates the heck out of things using chemical recognition among molecules, you now must adjust to the idea that the cell also tends to repeat itself. The cell sometimes has a backup plan.

Let's say that there's a specific pathway that turns Yellow into Blue. The steps of this process are Yellow is turned into Purple, which is turned into Red, which is then turned into Blue. Cellular processes often involve many steps like this to get from the initial to the final molecule.

At every step, a specific enzyme catalyzes each reaction. But what if the enzyme responsible for turning Purple to Red is faulty and can't recognize the Purple reactant? In some cases, the cell may have a second enzyme at the ready, one that enters the pathway at that point and takes over the metabolism of Purple to Red. Not all pathways have this redundancy, but it is fairly common.

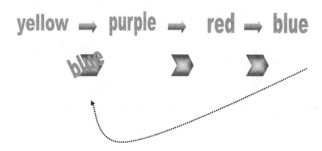

Negative feedback in a metabolic pathway.

Enzymes can't always be "on," either. At some point, the cell is going to have enough Blue and want the enzymes to stop making more. A common way for a cell to turn off an enzyme is by using some endpoint molecule in the pathway to do it.

In our made-up example, when the concentration of Blue hits a certain level, the pathway will shut down, as in the figure. Here's how: some of the Blue molecules attach to the first enzyme in the pathway, the one that acts on Yellow, and distort it. The distorted enzyme can no longer interact with Yellow, so the pathway shuts down. When Blue depletes, the pathway starts again.

Another way the cell can shut down is through a mechanism called competitive inhibition. If you were trying to catch a volleyball, it would be a lot harder to do if you already had a beach ball in your hands. Competitive inhibition operates much the same way. A molecule that doesn't actually activate the enzyme may park itself in the active site, inhibiting the enzyme's interaction with its real target.

Then there are the molecules that either activate or inhibit the enzyme by influencing its folding. These molecules, known as allosteric (*allo* = different; *steric* = shape) regulators, act away from the active site on the enzyme.

Imagine trying to push a square sponge into a round hole. You have to push the sponge on one side and then the other to conform it to the round shape. In much the same way, an allosteric activator will push on an enzyme at one site to get it into the right shape to recognize its target molecule. An allosteric inhibitor does the opposite: it changes the enzyme's shape so that the active site does not recognize the target.

You'll notice in our color example that the end product of the pathway (Blue) regulates the beginning enzyme of the pathway. It's far more efficient to flip off the pathway at its beginning. Doing so prevents the cell from making any unusable intermediate products (no unnecessary Purple or Red!) or wasting energy. Is nature always this efficient? Not always, but often!

The Least You Need to Know

- Spontaneous movement of molecules down a gradient releases free energy and is exergonic.

- The building of molecules requires input of energy and is an endergonic reaction.

- Coupled reactions involve the free energy from an exergonic reaction fueling an endergonic reaction.

- ATP carries potential energy in the third phosphate bond, which can be released to fuel endergonic processes.

- Enzymes help lower the activation energy required for the reactions of life so they can proceed.

- Enzymes use molecular shape recognition, ensuring that they target the right molecules and aiding in their regulation.

7

Respiration:
Energy Extraction

In This Chapter

- Harvesting energy from sugar
- Glycolysis in two stages
- The Krebs cycle happens twice per glucose
- The electron transport chain and an ATP payoff
- Fermentation and alternative pathways

Everything that goes on in a cell—and thus, everything that goes on in your body—requires energy in the form of work done. The initial source of energy for most life is the radiant Sun. Certain kinds of cells can capture this energy and transform it to package up molecules of sugar. Here, we examine how cells extract the energy from those Sun-sourced sugar molecules. Your job is to follow the energy. To make this easier, I've provided an ongoing energy tally using a new high-tech invention of mine, the Energy Tracker.

No O₂ Required: Glycolysis

The sugars involved in glycolysis are glucose molecules. The first step in extracting the energy in their bonds is to break this molecule into two pieces. Recall from the lysosome (the body that breaks things) that *lyse* means "break." It's also helpful to know that the root *glyco* means "sugar." Thus, glycolysis means "sugar breaking." For our cells to extract energy from glucose, they must first rearrange the molecule so that it has two equal halves. Then … snap!

Biohazard!

Don't confuse your respirations. Cellular respiration means something different from what we do with our lungs. The strict definition of cellular respiration is that it is the sum of all of the pathways, aerobic (with O_2) or anaerobic (without O_2), involved in breaking down organic molecules to build ATP.

Upon entering the cell, a glucose ends up in the cytoplasm, where this initial rearrangement and breakage occurs. Although the overall processes described here are referred to as cellular respiration, which calls to mind the exchange of O_2 and CO_2, the glycolysis step of glucose breakdown does *not* require oxygen. The processes of respiration that require oxygen are called aerobic; those that do not are anaerobic.

It does, however, require a lot of enzymes, a couple of initial ATP molecules for energy investment, and some molecules specialized to pick up and carry electrons around. In fact, as this chapter unfolds, you find that moving electrons around is one of the leitmotifs of energy extraction from glucose molecules. Another motif to watch for is the movement of phosphates—those little charged molecules play a major role throughout the process of energy extraction from sugar. And finally, keep an eye on those protons (H+).

This chapter is also packed with examples of redox reactions. When a molecule receives an electron, we refer to it as being reduced. When it loses an electron, we refer to it as being oxidized. Because electrons get shifted around several times in cellular respiration, added here, stripped away there, we refer to this series of electron transfers as redox reactions.

Investment and Payoff: Coupled Reactions

Glycolysis starts with a coupled reaction. The kickoff is the energy-investment phase, in which an enzyme called hexokinase strips the phosphate from an ATP and sticks it onto the glucose (1 in the following figure).

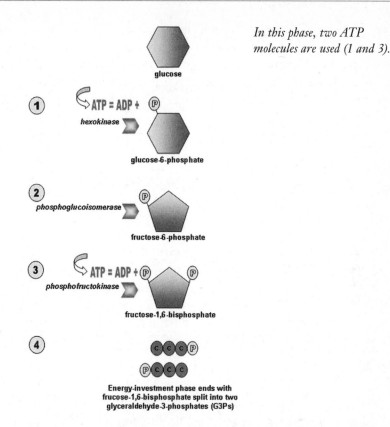

In this phase, two ATP molecules are used (1 and 3).

glucose

① ⌇ ATP = ADP + Ⓟ

hexokinase ➤

glucose-6-phosphate

② Ⓟ

phosphoglucoisomerase ➤

fructose-6-phosphate

③ ⌇ ATP = ADP + Ⓟ   Ⓟ

phosphofructokinase ➤

fructose-1,6-bisphosphate

④   ⒸⒸⒸⓅ
    ⓅⒸⒸⒸ

Energy-investment phase ends with frucose-1,6-bisphosphate split into two glyceraldehyde-3-phosphates (G3Ps)

The attachment of a phosphate to another molecule is called phosphorylation. The transfer of this phosphate group achieves two things: the sugar cannot leave the cell because the phosphate group is charged and can't get out, and the glucose is now also a more reactive molecule. The depleted ATP is now an ADP. It's our first coupled reaction! The free energy from the breakdown of ATP provides the energy required for hexokinase to phosphorylate glucose.

Energy Tracker

ATP count = –1

Rearranging the Molecular Furniture

You may be wondering why the cell bothers to attach the phosphate to the glucose. The final outcome here is to split the glucose molecule into two parts, so why doesn't the cell just go ahead and break the thing?

It's a matter of balance and fit. Each enzyme in the pathway to breakage can only recognize specific conformations of molecules, and glucose itself is not two balanced halves joined into a single molecule. Hexokinase must attach that first phosphate so that the next enzyme, phosphoglucoisomerase (see 2 in the figure), can rearrange the glucose into a different six-carbon sugar, fructose-6-phosphate. The next enzyme in the chain, phosphofructokinase (see 3 in the figure), as its name implies, adds a second phosphate onto this fructose, creating a molecule that has two halves that are mirror images of each other.

All of this molecular rearrangement to this point was solely to produce a molecule with two equal halves that can then be broken into two. The molecule is now ready to be "lysed," or broken into two parts.

Bio Basics

You'll notice enzymes called kinases act throughout glycolysis. The prefixes may differ—in glycolysis, we have hexokinase, phosphofructokinase, phosphoglycerokinase, and pyruvate kinase—they all do the same job: adding phosphates to molecules. The prefix usually describes the molecule receiving the phosphate. Thus, a hexokinase attaches a phosphate to a hexose, or six-carbon sugar, like glucose.

Phosphofructokinase is a kinase and required a phosphate to do its job in step 2. That phosphate came from a second ATP molecule, leaving behind another depleted ADP. Now we have had two coupled reactions involving energy investment from ATP, using two ATP molecules. The cell is currently at an energy deficit of two ATPs. But don't worry. The next steps of glycolysis make up for it and then some.

Energy Tracker

ATP count = –2

The energy-investment part of glycolysis ends with the splitting of the six-carbon fructose with the two phosphates on it to make two three-carbon sugars called glyceraldehyde-3-phosphates (G3P; the 3 carbon is where the phosphate is attached). Each G3P (4 in the figure) has a single phosphate. It's time for some energy payoff. Keep watching the phosphates, but also monitor the bouncing electrons and portable protons.

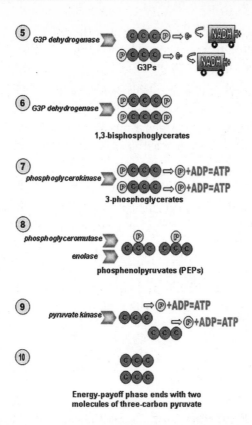

This part of glycolysis produces 2 NADH electron carriers and 4 ATPs.

Our next step in rearranging the molecular furniture involves stripping a *hydride* from each G3P (5 in the figure). Recall that removing electrons is oxidation (electron transfer alert!), so G3P is our first molecule to officially undergo oxidation. That electron has to go somewhere. It happens

def•i•ni•tion

A **hydride** consists of two electrons and one proton, or a hydrogen atom plus one extra electron.

that there is a handy electron transport molecule hanging out nearby. This molecule, called nicotinamide adenine dinucleotide (which we will simply call NAD+, like everyone else), serves as an electron truck, transporting the electrons it picks up to a different part of the cell. Because we have two G3P molecules, we have two NAD+ molecules there to collect the released electrons.

Redox alert! As you may have noticed, NAD+ just picked up electrons. That means that NAD+ is reduced, receiving electrons and becoming NADH. Those electrons came from hydrogen, so there's a proton left over now, designated as H+. The proton will trail around after the NADH (step 5 in the figure), perhaps keeping an eye on its lost electron. Your job now is to keep an eye on the fate of the electrons, now associated with NADH, and the trailing proton.

That electron transfer was exergonic as the electrons settled into a more stable conformation in NADH. Hmmm … wonder what the cell can do with all that free energy? How about slapping a phosphate onto something? When in doubt about the next step, just think "Oh, let's move a phosphate somewhere."

In this case, the phosphate is attached to the now oxidized G3P molecule, turning it into 1,3-bisphosphoglycerate (6 in the figure). It's "1,3" because it now has a phosphate on its 1 carbon and another on its 3 carbon. Don't forget that we actually have two of these 1,3-bisphosphoglycerates because we made each of them from the two G3Ps.

The enzyme responsible for attaching the phosphate in this step is glyceraldehyde-3-phosphate dehydrogenase. As its name implies, it also "dehydrogenizes" the G3P, handing the electron over to NAD+ and leaving the H+. Thus, it handles both the exergonic and the endergonic steps in this coupled reaction.

Energy Tracker

NADH count = +2 (and trailing protons!)

ATP count = −2

What happens next? Remember, when in doubt, just think, "Oh, let's move a phosphate somewhere." That "somewhere," in this case, is ATP. Yep, it's true. Phosphoglycerokinase (kinase alert!) strips a phosphate from our 1,3 bisphosphoglycerate and adds it to … an ADP. Et voilá! We have ATP (7 in the figure). Do this twice (one for each 1,3 bisphosphoglycerate we're working with), and we get two ATPs out of this step!

Energy Tracker

NADH count = +2 (and trailing protons!)

ATP count = +2 + −2 = net of zero

We've just broken even on our energy investment/payout. Better than the stock market's been doing lately.

Onward. We're almost there. Can you believe that the cell goes to all of this trouble just to break a glucose molecule in half? And this is only the first of three processes.

Remember that we just took a phosphate off of our 1,3 bisphosphoglycerate and gave it to an ADP. That left us with two molecules of 3-phosphoglycerate (step 7 in the figure). As its name implies, there's a phosphate on that 3 carbon. Hmmm ... wonder what we should do next? Oh, let's move a phosphate somewhere.

The cell is being incremental at this point. First, an enzyme called phosphoglycero-mutase moves a phosphate around. Then an enzyme called enolase does some fancy footwork with water and carbon bonds. The outcome of all this shifting around is two molecules of PEP, or phosphoenolpyruvate (8 in the figure).

Gee ... what should we do with the PEPs? Oh, let's move a phosphate somewhere. Enter the final kinase of glycolysis, pyruvate kinase. It grabs a phosphate off of the second carbon of each of our two PEPs and attaches each one to a lurking ADP (9 in the figure). Et voilá! Two more ATPs! Finally—some net return on all of this investment.

> **Energy Tracker**
>
> NADH = 2 (and trailing protons!)
>
> ATP = +4 + –2 = 2 (net for glycolysis: +2 ATPs)

What do we have left? Two lovely molecules of three-carbon *pyruvate* (10 in the figure), derived from our starting material of a single six-carbon glucose. Next time you eat, just think of all that your body has to do to get at that energy in your food. And we've only just begun. Now it's onward to the mitochondria.

To the Mitochondrion: Krebs Cycle

The events to this point didn't require oxygen. That may be confusing given that we called part of that process "oxidation," but remember that the term simply refers to the extraction of electrons from a molecule. Now however, the presence or absence of oxygen matters. If it is present, the two molecules of pyruvate (also called pyruvic acid) just produced from a single glucose molecule move into the mitochondrion for further processing and energy extraction, beginning with the Krebs cycle.

O_2 Required ... CO_2 Released ...

Now is the time to recall your mitochondrial structure from Chapter 5. The mito-chondrion is the site of steps two and three of energy extraction from glucose. Step 2 occurs in the matrix, or central part of the mitochondrion, and step 3 occurs primar-ily along the inner mitochondrial membrane.

This second step is a cycle because the process begins with a specific molecule, cycles through some structural changes, and in the final step again produces the starting molecule. But the cell must first do a few things with those pyruvates.

The Krebs cycle starts with the product of glycolysis, pyruvate. Each of its three carbons ends up in a carbon dioxide by the end of the cycle. During each cycle, 4 NADH molecules are formed, as are a single FADH$_2$ and an ATP. This cycle happens twice for each glucose, once for each of the two pyruvates.

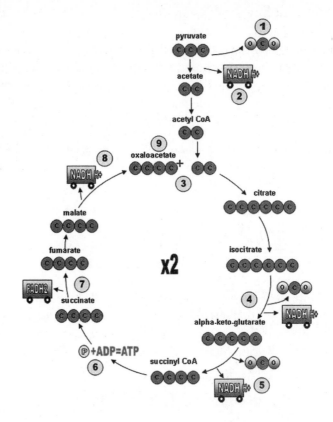

Remember that we have two pyruvates from glycolysis of a single glucose, and they each have three carbons. Just about the second each pyruvate enters the mitochondrion, however, it loses a carbon in the first molecule of carbon dioxide released in this "respiration" process (1 in the figure). Aw, it's our first CO_2.

Organic Molecules Exhausted

Simple math tells us that 3 − 1 = 2 carbons left per pyruvate. As we step through what happens next, you'll notice that each of the remaining carbons also ends up in a molecule of CO_2. Thus, the fate of each of the three pyruvate carbons is to participate in forming CO_2. The increasingly shrinking pyruvate molecule undergoes further lightening as NAD+ assists in its oxidation (electron transfer alert!), picking up a couple of electrons from it (2 in the figure). The resulting molecule is called acetate.

Energy Tracker

NADH = 2 + 2 = 4 (and trailing protons!)

ATP = +4 + –2 = 2 from glycolysis

Acetate now gets to take its vitamins. A molecule of coenzyme A (CoA), made from a B vitamin, latches onto the acetate, but it doesn't get a very firm grip. The result is an acetyl CoA, highly unstable because of that edgy, high-energy bond between CoA and acetate. The bond is perched on the edge of an energy cliff, ready to transfer that packed-up burst of potential energy to whatever reaction needs it. It's time for a cycle.

The unstable acetyl CoA feeds into the Krebs cycle. First, the acetyl part of acetyl CoA is added to oxaloacetate (3 in the figure) to make citric acid (also called citrate). Watch later for oxaloacetate to show up again.

After a rearrangement step, another molecule undergoes another oxidation (electron transfer alert!), with NAD+ picking up the electrons to become NADH (4 in the figure). At the same time, one of the two remaining carbons from that original pyruvate joins two oxygens to produce another CO_2. Remember that this process involves each of the two pyruvates derived from our original glucose, so for our energy tracker of one glucose, we're doubling our numbers.

Energy Tracker

NADH = 2 + 2 + 2 (1 per pyruvate) = 6

ATP = +4 + –2 = 2 from glycolysis

In the next step, that third and final carbon from pyruvate joins itself into a CO_2, also allowing for another oxidation reaction to produce another NADH (5 in the figure). We've now used all of those carbons from the pyruvates and yielded a total of six carbon dioxides per glucose during the Krebs cycle. The organic molecules have been exhausted. You're probably just as exhausted having stepped through it all with them.

Energy Tracker

NADH = 2 + 2 + 2 + 2 (1 per pyruvate) = 8

ATP = +4 + –2 = 2 from glycolysis

You = depleted?

Yet, there's still a bit more energy to transfer. Another step of the cycle takes care of that CoA, which gets the boot for the sake of a phosphate group. The phosphate group ultimately ends up on our favorite energy-packing molecule, ATP (6 in the figure). Remembering that this happens to each pyruvate that enters the cycle, giving us two ATPs derived in the citric acid cycle from one glucose.

Energy Tracker

NADH = 2 + 2 + 2 + 2 (1 per pyruvate) = 8

ATP = +4 + –2 = 2 from glycolysis + 2 from Krebs = 4

We're almost done with the cycle. In steps near the end, we meet a new electron carrier, flavin adenine dinucleotide, or FAD. During yet another oxidation reaction (electron transfer alert!), FAD becomes $FADH_2$ (7 in the figure). Two steps later, another oxidation reaction yields the final energy-carrying molecule of the citric acid cycle, another NADH (8 in the figure).

Energy Tracker

NADH = 2 + 2 + 2 + 2 + 2 (1 per pyruvate) = 10 (2 from glycolysis and 8 from citric acid cycle)

$FADH_2$ = 2

ATP = +4 + –2 = 2 from glycolysis + 2 from Krebs = 4

Although we haven't gotten into the molecular details, in this process, we've worked our way from a starting molecule of oxaloacetate through seven intermediates that have been oxidized and rearranged. The seventh intermediate, malate, lost its electrons to NAD+ as described previously and is metabolized into oxaloacetate again (9 in the figure). The cycle has ended. Or ... did it just begin again?

To the Inner Membrane: Building ATP

Wow. We just went from producing a couple of ATPs and NADH molecules in glycolysis to what must look like a great energy payoff during the Krebs cycle. Our total score now for our single glucose molecule is 10 electron-toting NADH, 4 ATPs, and 2 $FADH_2$. But the energy-packaging process hasn't even really gotten fired up yet. We're going to follow our electron carriers, NADH and $FADH_2$, as they head for the inner mitochondrial membrane with their precious electron cargo—and those protons (H+) that keep following them around. It's time for the Mother of All Coupled Reactions.

Bouncing Electrons, Flowing Protons

Stuck in the inner mitochondrial membrane like so many peanuts trapped in chocolate is a series of Very Important Proteins associated with a few electron carriers of their own. They serve as the drop-off point for those electrons that NADH and FADH$_2$ have been trucking around the cell all this time.

Bio Bits
If prokaryotes do not have mitochondria, do they "do" cellular respiration? The answer is, yes, but like any mom-and-pop operation, they just do it out in the middle of the room (cytoplasm) instead of having a special area of the factory dedicated to it. (See Chapter 5 for a refresher on prokaryotes.)

You can imagine this chain as stair steps. If you took a soccer ball and balanced it at the top of a flight of stairs, it would have quite a lot of potential energy in that unstable position. Pretend you're NADH and that ball is one of the electrons you're carrying. You ditch it by rolling it to the step just below you. What does the ball do? The same thing the electron does in that chain of proteins in the inner mitochondrial membrane: It bounces down, step by step, until it reaches its most stable and least-energized state. In the case of the ball, that's the floor. In the case of the electron, a different fate awaits.

As that electron bounces down the chain, which we naturally have named the electron transport chain, its energy transfers bit by bit. This step-by-step transfer results in the harvesting of greater amounts of useful energy than the cell would get if it just dropped the electron in one huge fall from its high-energy position.

The gently bouncing electron releases energy that the proteins in the chain grab. These Very Important Proteins use the energy to pump protons against their will across the inner membrane into the space between the two membranes, the inter-membrane space.

Remember those protons (H+) lingering around the reduced NADH molecules (1 in the figure)? This is their fate—to be pumped against their gradient into the inter-membrane space (2 in the figure). There are already relatively more protons in this space than there are in the matrix, yet the cell uses this precious electron-transfer energy to pump in more. Hmmm ... wonder why that is? You can see it coming, can't you? It's another coupled reaction!

The NADH and FADH₂ electron carriers drop off electrons at the electron transport chain (1). The electrons fall down the chain, providing the free energy to pump the H+ across the membrane (2). The protons exit through the ATP synthase (3), which uses this energy to add a phosphate to ADP to make ATP (4).

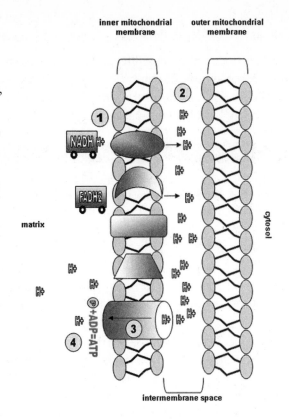

If you couldn't see that coming, that's OK. I'll explain it right now.

The Big Payoff: ATPs in Abundance

Those protons packed against their concentration gradient into the intermembrane space are like any frantic, agitated crowd: Their goal is to find a way out. Naturally, the cell provides one in the form of a protein with a special channel that happens to be perfectly made for protons (3 in the preceding figure). When the protons find this exit, they use it, flowing downhill—energy-transfer alert!—from high to low concentration, from the intermembrane space to the matrix. They're "falling" down their gradient and in doing so, they release free energy. It's an exergonic reaction just looking for a nice endergonic reaction to settle down with.

Lurking on the matrix side of that protein are many molecules of ADP and a bunch of phosphates. As the protons pass through the protein channel, the energy that their passage releases transfers to the endergonic process of adding a phosphate to each of these ADPs (4 in the figure). This process goes by the fancy name of substrate-level phosphorylation and yields the bulk of ATPs from aerobic respiration.

The process requires about 32 to 34 ADPs, to be kind of exact. You may not be surprised to learn that the protein that does the attaching while allowing protons to flow through is called ATP synthase: the enzyme that synthesizes ATP. The cell finally has its biggest payoff: 32 to 34 energy-packed ATPs.

Bio Bits

The yield of final ATPs (36–38) is theoretical, and the reality can vary. NADH affects the final count because it must transfer its electrons to a carrier on the inside of the mitochondrial membrane. If NAD+ receives them, the yield is about 3 ATPs per carrier, but if FAD takes them, the yield is only 2. Different tissues use different carriers. The brain, for example, yields fewer ATPs because it uses FAD.

Energy Tracker

NADH, $FADH_2$ = depleted

ATP = +4 + –2 = 2 from glycolysis + 2 from Krebs + 32 to 34 from electron transport chain = 36 to 38 total from a single glucose molecule!

If you're still following the bouncing electron, you may be wondering what its final fate is. Lurking at the end of the electron transport chain are protons and oxygen. When the electrons tumble down into their lower-energy state, they join the protons and the oxygen to make a familiar molecule: H_2O, or water. Thus, oxygen is the final electron acceptor of cellular respiration. We started back at glycolysis with a glucose and ended up making some carbon dioxide, water, and a whole lot of ATP.

Alternative Pathways

Without oxygen, the cell still has an alternative: fermentation. In your muscles, your muscle cells turn to lactic acid fermentation in which NAD+ simply reduces the pyruvates, yielding lactate, or lactic acid, as a product. This process does not yield any CO_2, and all you get out of it is those two ATPs from glycolysis. And then there are yeasts, which turn to a different kind of fermentation.

When live yeast is mixed into bread dough, no oxygen reaches the yeast, so these single-celled eukaryotes turn to two-step alcohol fermentation. Its first step is reminiscent of oxygen-based respiration: in both processes, the pyruvates first release a carbon to form CO_2. The two-carbon acetylaldehyde that remains in alcohol fermentation then loses electrons to NADH to become ethanol.

In the rising bread dough, the yeasts produce the CO_2, which puffs up the dough, causing it to rise. The little fungi—yeasts are fungi—are also busily making ethanol, which will "burn off" when you bake the bread.

The Least You Need to Know

♦ Glycolysis occurs in the cytoplasm in energy-investment and energy-payoff phases, yielding two net ATPs and two NADH.

♦ If oxygen is present, the Krebs cycle cycles through seven intermediates in the mitochondrial matrix and yields eight NADH, two $FADH_2$, two ATPs, and six carbon dioxides per glucose.

♦ The electron transport chain uses the energy of falling electrons from NADH/$FADH_2$ to pump protons into the mitochondrial intermembrane space against the gradient.

♦ The protons pass down their concentration gradient through ATP synthase, providing free energy to link phosphates to ADPs.

♦ The final electron acceptor in aerobic respiration is oxygen, which combines with the depleted electrons to form water.

♦ The final ATP count from a single glucose in aerobic respiration is 36–38, compared to 2 from fermentation.

Chapter 8

Photosynthesis: Building Carbs

In This Chapter

◆ Sunlight and life

◆ Chloroplasts and chlorophyll in photosynthesis

◆ The light-dependent reactions

◆ The light-independent (dark) reactions

There's a famous observation that green chlorophyll is the only pigment we can see from space. Whether or not that's technically true, we can say that almost all life on Earth exists because of the power of this pigment to absorb the Sun's energy so certain organisms can use that energy to build sugar. You will be hard pressed to find an ecosystem on Earth that is not linked to the Sun through this process, called photosynthesis.

But wait, you may be thinking. Sunlight can't penetrate everywhere on Earth. How can almost every organism be linked to photosynthesis? Photosynthesis starts at the earth's surface, but the organisms that photosynthesize then contribute to the transfer of energy to other organisms, usually by becoming food for them. So even organisms at the depths of the ocean rely on events at the surface for their existence. And it all starts with a pigment's ability to soak up the Sun's rays.

Chloroplasts Capture Light

You may have noticed that leaves are generally green. They get their color from a pigment called chlorophyll, which reflects light waves in the green part of the color spectrum but absorbs light at some other important wavelengths. When the chlorophyll absorbs sunlight, that light energy transfers to electrons and kicks them up to a high energy state.

As you must now realize, what goes up in nature must come down. And it is during the electron's fall that the plant captures energy and uses it to build energy-carrying molecules.

The plant carries its chlorophyll in those organelles called chloroplasts. Chloroplasts occur in a special part of leaves called mesophyll, where carbon dioxide enters and oxygen escapes. These gases enter and exit through pores called stomata (singular: stoma).

> **Bio Bits**
>
> Why do leaves change color in the fall? The pigments for brown, red, and yellow are always there, but chlorophyll masks them. With the fall season, physiological triggers of temperature and day length set off chlorophyll breakdown. As the green disappears, the other colors can shine through.

Chloroplasts, as you may remember, have a double membrane. Inside the chloroplast is a network of membranous sacs called thylakoids that are stacked like pancakes (called grana). Chlorophyll sits in the thylakoid membrane along with its cohort of proteins and other molecules required to grab sunlight energy and build energy-carrying molecules.

> **Bio Bits**
>
> Cyanobacteria, which are photosynthesizing prokaryotes, do not have chloroplasts. Instead, photosynthesis takes place in folds of the cell membrane, also called thylakoids.
>
> Plenty of nonplant organisms photosynthesize, including algae and several kinds of bacteria, like cyanobacteria, which contribute significantly to the photosynthetic activity on Earth. Organisms that do not photosynthesize include all animals, the Archaea (so far as we know), and fungi.

So what's going on in those chloroplasts that's so important? Oh, only the great circle of life. Biology books always give the following equation to illustrate what many consider the real great circle of life (the double-headed arrow means the reaction can go either way):

$$sunlight + H_2O + CO_2 \leftrightarrow C_6H_{12}O_6 + O_2$$

Or, in words:

Sun's energy + water + carbon dioxide $\leftrightarrow$ glucose and oxygen

Bio Basics _____

Scientists think that chloroplasts, like mitochondria, may once have been free living. In fact, evidence suggests that the ancestral cells that became the chloroplast were likely some kind of cyanobacteria.

Where have we seen glucose and oxygen lately? Yep, back there in cellular respiration, Chapter 7. Recall that cellular respiration by-products are water and carbon dioxide. The carbon dioxide forms during the Krebs cycle, and the water forms when depleted electrons from the electron transport chain join oxygen and protons.

And now you see that those very by-products of cellular respiration are used in photosynthesis to build the sugar and release the oxygen that cells then use in cellular respiration. That's why I call it the circle of life. And the circle, if it has a beginning, starts with sunlight and water.

Bio Bits
Although almost all life on Earth is tied to photosynthesis, there are exceptions. Deep in the earth is a community of bacteria that is completely disconnected from the Sun, using uranium as an energy source. Then there's the world's loneliest microbe, dubbed *Desulforudis audaxviator*, which translates loosely as "boldly traveling, sulfur-using rod." These bacteria live two miles beneath the earth's surface and are the only known sole occupants of any ecosystem.

More Light! The "Light" Reactions

Photosynthesis (*photo* = light) occurs in two stages. Because the initial energy-input reactions require sunlight, we call them the light-dependent (or light) reactions. The second set of reactions can happen in light or not, so we call them the light-independent (or dark) reactions. Photosynthetic organisms use the light reactions to build energy-carrying molecules. They use the dark reactions to harvest this energy-packaging carbon into sugar molecules.

The Electromagnetic Spectrum

Remember how we followed the electrons around during cellular respiration? You're going to follow them around in photosynthesis, too. In photosynthesis, the origin of these electrons is the splitting of water. But before electrons can start moving at all, they must first be energized.

def•i•ni•tion

A **photon** is a packet of light energy that is also a wave with properties of frequency and wavelength. The different wavelengths form a spectrum called the **electromagnetic spectrum**. We see wavelengths at the center of the spectrum as visible light when they are reflected.

The initial source of energy is light from the Sun in the form of electromagnetic waves. Light is not only a wave, it is also particulate. The packets of energy that make up the wave are called *photons*, which travel at different frequencies.

These different frequencies or wavelengths make up the *electromagnetic spectrum*. Although the spectrum spans wavelengths from the high-frequency gamma waves that can break DNA to low-frequency radio waves, the part that we see is the visible spectrum.

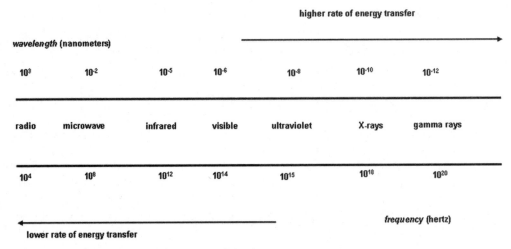

The range of wavelengths and frequencies of the electromagnetic spectrum.

We see the pigment chlorophyll as green, which means that it absorbs other wavelengths of light in the color spectrum but reflects much of the green. The important part for photosynthesis is what the eye doesn't see: the light energy that the chlorophyll absorbs—the light energy that transfers to electrons and gets them jumpy and excited, kicking them up to a perilously high energy level so that they have nowhere to go but down.

Photosystems II and I

This absorption of sunlight takes place in chlorophylls located in a pair of specialized molecular systems. Two systems are required to complete the light reactions. The first system in the series is Photosystem II (PsII), and the second in the pair is Photosystem I (PsI). Yep, that's confusing. The reason for the backward numbering is that researchers discovered and named PsI before discovering and naming PsII, and PsI evolved first.

Each photosystem contains a light-harvesting complex, a reaction-center complex, and a primary electron acceptor. The light-harvesting complex is responsible for, um, harvesting the sunlight. It consists of accessory pigment molecules, including chlorophyll and some others called carotenoids. These pigments absorb light that energizes their electrons, resonating their energy inward to a special pair of pigment molecules lying at the center of it all: the reaction center. The reaction center then boots the electrons to the primary electron acceptor.

The transfer to the reaction center occurs through electrons being excited in a series, rising and falling and exciting the next electron in the chain. At the reaction center sit two special chlorophylls, named chlorophyll *a*. These chlorophyll *a* molecules in PsII have been designated as P680 because they absorb at a wavelength of 680nm (nanometers). This absorbed light gets the electrons in the chlorophyll excited and jumpy.

Biohazard!

Don't think that the term "accessory pigments" means that the nonchlorophyll *a* pigments are unimportant. The P680 reaction center is pretty limited in what it can absorb. Accessory pigments make up for that limitation. Because carotenoids and chlorophyll *b* (or *c* or *d* in algae) absorb light in different parts of the spectrum, they broaden the possibilities for soaking up light energy and can also help dissipate excess energy.

The superexcited P680 electrons from the P680 chlorophyll *a* pair pass to a unit called the primary electron acceptor. Each P680, now down by one electron, becomes P680+ and is ready to receive another electron to transfer.

P680 has a reputation as the strongest biological oxidizing agent known. When water is split into two electrons, two protons, and an oxygen atom (shades of the final step of cellular respiration in reverse), each P680+ in the paired chlorophyll *a* molecules immediately replenishes itself by grabbing an available electron.

> **Bio Basics**
>
> It's important to remember that photosynthesis begins with electrons that are derived from the splitting of water that then become excited by energy from the Sun.

As you can see if you reread the previous paragraph, without water, plant photosynthesis does not happen. No water, no P680+ electron source, no photosynthesis. Without water, life on Earth, relying on photosynthesis as it now does, could not continue.

Once each P680 gives up its electron to the primary electron acceptor and becomes P680+, those transferred electrons still have quite a journey ahead of them. After all, I promised you some energy-carrying molecules, and we haven't even made any of those yet.

> **Bio Bits**
>
> Some organisms begin photosynthesis with the splitting of hydrogen sulfide or plain old molecular hydrogen instead of water. These bacteria probably are exhibiting an ancient version of photosynthesis. Hydrogen sulfide is a deadly gas, however, and the splitting of water is healthier for current life on Earth, especially us O_2 breathers.

Products of the Light Reactions

The primary electron acceptor passes each electron into an electron transport chain. Yes, this is very much like the electron transport chain we encountered in cellular respiration. And the outcome is similar. As each electron falls down this chain, proteins use this energy to push protons (some from the splitting of water) against their gradient across the thylakoid membrane into the thylakoid space (the inside of the pancake).

The protons "want" to cross back out of the thylakoid, into the thick, liquidy stroma of the chloroplast. And they do so, just as in cellular respiration, by passing through an ATP synthase, which couples the free energy to the energy-requiring job of adding a phosphate to ADP to make ATP. The result is a single ATP. The chloroplast is no mitochondrion in this regard.

Yet, even when they finish this electron transport chain, these electrons—the same pair of particles that popped out of PsII—are not quite done. They must now "land" in the reaction center of PsI, where a couple more chlorophyll *a* molecules await, this time designated P700 because that's the wavelength they absorb. Light strikes the light-harvesting center of PsI, the energy wave travels to the reaction center, and those original electrons are excited again and transferred to yet another primary electron acceptor. Each P700 is now P700+ until the next depleted electrons from PsII arrive.

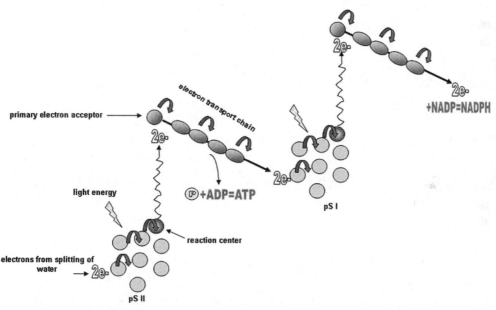

Energized electrons bounce through two photosystems and two electron transport chains to yield ATP and the plant version of an electron carrier, NADPH.

Meanwhile, our well-traveled electrons make their way through a second electron transport chain, but this one has a twist. There are no protons involved, no gradient, and no ATP synthase. Instead, our worn-out electrons bounce down that chain to their final, if temporary, rest in the arms of NADP+, the plant's version of NAD+. To be completely reduced, NAD+ must pick up both electrons.

The resulting molecule, NADPH, is the second energy-carrying molecule to emerge from the light reactions. The electrons it picked up, even after all of that falling, are still at a higher energy level than they were when they were pulled off of that split water molecule at the beginning of the light reactions.

You may have noticed that the light reactions are all about repeats. Two photosystems. Two light-harvesting centers. Two reaction centers. Two primary acceptors. And two electron transport chains. After all of that work, the plant now has chemical energy in an ATP and reducing power in an NADPH to send onward to the dark reactions.

No Light! The "Dark" Reactions

The ATP and NADPH the plant just made now get to do their jobs in the processes known as the "dark" reactions. It doesn't have to be dark for these reactions to take place, but no light is required, either. Now the work of grabbing, or fixing, carbon from carbon dioxide and building a nice three-carbon sugar can begin.

Wait—did I just say three-carbon sugar? But glucose has six carbons, right? It's true. When the dark reactions are complete, the final product is not glucose, but a smaller sugar called glyceraldehyde-3-phosphate, or G3P. Glue a couple of them together, and you'll get your glucose.

To get to G3P, the plant uses a cycle. You may have noticed nature's tendency to recycle, and this is no exception. Just as we saw with the Krebs cycle in cellular respiration (see Chapter 7), this cycle starts with a certain molecule and ends with that same molecule. But a lot sure does happen in between.

A Cycle of Carbons

This light-independent or dark reaction cycle is the Calvin cycle. It starts with a five-carbon sugar called ribulose bisphosphate (or RuBP, which we say "Ru-bee Pea"), which has a phosphate attached to each end (hence the *bis*, for two).

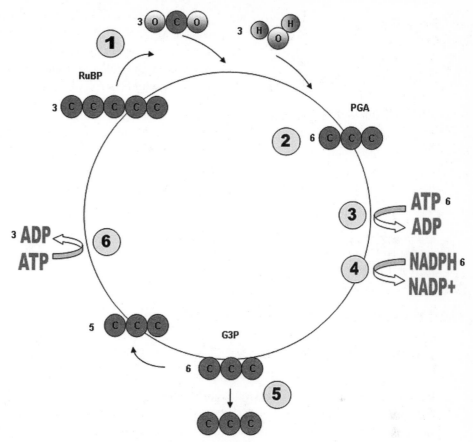

This figure sums the events of the three Calvin cycles required to make a single molecule of G3P. It begins and ends with five-carbon RuBP.

Here's where I hope you're up on your multiplication tables for multiples of three. To make one three-carbon G3P molecule, the cycle has to complete three times. So for one G3P molecule, the plant will use three RuBPs. That's a total of three five-carbon molecules as the starting material, or 15 carbons. In this cycle, the plant acquires three more carbons from carbon dioxide, for a total of 18. By the end, three carbons are used in G3P, and the other 15 return to the cycle.

More Rearranging of the Molecular Furniture

The first step is for the plant to grab a carbon dioxide and attach the carbon to RuBP (1 in the previous figure). For one G3P, the plant must do this fixation step in each of three cycles, using an enzyme called RuBP carboxylase, or rubisco for short. Rubisco enjoys a reputation as the most abundant protein on Earth in addition to having a name that sounds like a snack food. In each cycle, the molecule that results from the fixed carbon is an unstable six-carbon sugar. We have three of these total from the three cycles, so our carbon count is now 18 carbons.

This unstable molecule quickly splits into two three-carbon molecules, called 3-*phosphoglycerate* (PGA), each with one phosphate attached (2 in the figure). The next step is to add a phosphate to the other end of each of these molecules, forming two 1,3 bisphosphoglycerate molecules. Guess where those phosphates come from? Yep, from that ATP the plant made in the light reactions (3 in the figure).

The plant needs a total of six ATPs for this step. In the three Calvin cycles required to make one G3P, it will need to make six 1,3 bisphosphoglycerates and thus needs six phosphates. Note that with six 1,3 bisphosphoglycerates, there are still 18 carbons involved for one G3P.

The Result: Sugar (and Other Things)

Speaking of G3P, it's finally time to make some. Enter NADPH (4 in the figure). It reduces the 1,3 bisphosphoglycerates—electron transfer alert!—and the sugar at the same time gives up a phosphate. For making one G3P, six NADPH molecules will transfer electrons in this reduction step, and six 1,3 bisphosphoglycerates will give up their phosphates. The result? Six three-carbon G3P molecules. We still have 18 carbons. Remember that when we started, we had 15 from the three starter RuBPs and fixed three from three molecules of CO_2.

Now for the payoff. In the next step, one G3P is reserved for building glucose (5 in the figure) or other useful biological molecules. We had six G3P molecules, so guess what the fate of the remaining five is? Well, they go on to another step in which the 15 carbons in the six three-carbon G3Ps are rearranged into three molecules of the five-carbon RuBP. In this regeneration process, three more ATP molecules give up their phosphates (6 in the figure) so that RuBP is truly a *bis*phosphate again. Yes, we are now back where we started, and the three cycles required to make one G3P are complete.

To make that single G3P, the plant had to produce and use nine ATPs and six NADPH molecules. Think about the electrons and where they started. The plant obtains them from split water, passes them through two photosystems onto NADPH, which in turn uses them to reduce 1,3 bisphosphoglycerate and produce G3P. The G3P is the ultimate fate of those well-traveled electrons.

The G3P heads to other pathways where it might be used to build glucose or some other kinds of carbohydrates.

The Least You Need to Know

♦ Photosynthesis = sunlight + water + carbon dioxide to build sugars in chloroplasts and bacterial membranes.

♦ The usual electron source for plant photosynthesis is the splitting of water.

♦ Photosynthesis is divided into light-dependent (light) and light-independent (dark) reactions.

♦ In the light reactions, two photosystems, each with a light-harvesting complex, reaction center, and a primary electron acceptor, use pigments to harvest the Sun's energy to excite electrons.

♦ Electrons fall down electron transport chains, with the free energy used to build ATP and reduce NADP to NADPH.

♦ The dark reactions use ATP and NADPH and carbon fixation to produce a three-carbon sugar in the Calvin cycle.

Cell Communication

In This Chapter

- ◆ Why cells communicate
- ◆ First and second messengers
- ◆ Passing along the message inside the cell
- ◆ The response to the message
- ◆ Cell suicide

Considering that your body consists of cells by the trillion, it may not surprise you to learn that these cells must talk to one another. Cells achieve communication primarily through a chemical language. The messengers that bring the molecular messages can either enter the cell directly for delivery or talk to other molecules at the cell membrane doorstep, which in turn deliver the message on the inside. Every step of this process requires proteins—chains and chains of proteins in some cases, each passing along the message in a giant cellular game of Telephone. Let's take a look at how cells talk to each other and what they talk about.

What Do Cells Talk About?

Cells have much the same things to talk about as organisms do. A great deal of their conversation involves reproduction, either sexual or asexual, or at least some exchange of genetic information. They also discuss working together in colonies or in tissues, and often send messages to one another across long distances, for example from your brain to your big toe. (Wiggle your big toe. You just sent a message.)

These messages are often chemical but can be electric, as happens in nerve cells. Regardless of how it's sent or what it means, every message involves three stages. These stages—reception, transduction, and response—are very much like what you do when you yourself get a message from another person. You receive it, you process it, and you respond. Here's how cells help you do that.

How Do Cells Talk? Chemically.

Some cells talk to each other through direct and constant contact with each other. Others talk only through contact between proteins extending out of their cell membranes, while some cells communicate by sending chemicals out that will talk with cells they cannot touch.

Cells can touch to form channels of communication in a few different ways. Some cells, such as those in your heart, literally do form channels that run through the membrane of one cell and connect with a channel through the membrane of the adjacent cell. These tunnels through adjoining cell membranes, called gap junctions, allow cytoplasmic interchange between cells and let rapid communication happen. You can imagine why rapid communication would be important for heart cells that must contract at the same time.

Types of Interaction Between Adjacent Cells

Junction	Features	Tissue types
Gap junction	Connecting channels	Heart
Tight junction	Tightly sewn membranes	Bladder
Plasmodesmata	Plant tight junctions	Plants
Glycoproteins	Specific recognition	Many cell types

Plants have a similar kind of junction between cell walls that allows the flow of cytosol between cells. You will find throughout this book that plants have different names from animals for what are very similar things. We discovered this in Chapter 1 with plants having Divisions while everything else has Phyla. Well, plants also have plasmodesmata, rather than gap junctions.

Another form of junction between adjacent cells is the tight junction. This junction arises thanks to proteins that thread the cells together so tightly that liquid cannot flow between them. Your bladder is a good example of how important a junction like this is—that's one place where you want the fluid to stay in.

A less close association between adjacent cells happens when the proteins poking out of the membranes talk to each other. These proteins usually have sugar trees attached (thus, they are glycoproteins), shaped to recognize and interact with specific sugar trees poking out of other cells.

Sometimes, these trees help cells attach to each other, as when a fertilized egg reaches the uterine lining in a mammal. The egg doesn't just stick onto the lining with some kind of magical glue—instead, there are glycoproteins on the egg that recognize and interact with glycoproteins on the uterine wall. A similar kind of recognition is involved when the sperm fertilizes the egg.

These types of communication between cells all involve direct touching in different degrees. But often, cells must talk to one another over a distance, either long or short. To achieve this, they use specific chemicals for specific communication situations. If the message is "Grow, grow!", for example, the brain will send chemical signals called growth factors to the target cells to tell them to divide and make more cells.

Senders and Receivers

In the above scenario, brain cells send out chemical messages to other cells in the body. This kind of long-distance signaling involves the blood-borne messengers called hormones. For the receivers to receive this message, they must produce the exactly correct proteins to recognize the signal.

It's a bit like using different languages. If the brain sends a message in chemical French, the cells that receive the message must understand chemical French. Cells that make proteins that read only chemical Mandarin will not get the message.

In this way, an organism regulates the sending and receiving of messages and ensures that cells that are not the intended recipients do not accidentally recognize and respond to a signal inappropriately. If heart cells recognized messages intended to tell bone cells to divide, the outcome would not be good.

The long-distance signaling that hormones participate in is called endocrine or hormonal signaling. Some messages don't have to travel as far and may simply travel to the closest cell. This short-distance signaling is known as paracrine signaling (*para* = near or next to). Sometimes, a cell can even signal to itself using autocrine signaling.

> **Biohazard!** _____
>
> Don't let this focus on chemical signaling fool you. Not all cell communication is chemical, and some signaling mixes the electrical and chemical. A special case is synaptic signaling between nerve cells. The space between nerve cells is called the synapse. A nerve cell releases a chemical called a neurotransmitter into the gap between the cells. These molecules spread across the gap to the opposite nerve cell and bind receptors on its membrane, sending this cell the message to fire off—or not—an electrical signal. As you may imagine, this entire process happens at blinding speed.

Direct Messaging: Inside the Cell

However far the messages must travel, they can reach the inside of the target cell in one of two ways. The messenger itself can enter the cell and deliver it, or the messenger can have a chat with a protein at the cell entrance and pass the message along. In either case, the initial messenger is called the first messenger because it is the first chemical to carry the message to the cell.

First messengers that signal directly inside the cell are usually lipid-based hormones, such as testosterone and estrogen, or molecules that enter through specific membrane proteins, as does thyroid hormone.

def•i•ni•tion _____

The process of transcription involves making a copy of DNA using RNA as the copy molecule. The proteins involved in kicking the process off are often in a class called **transcription factors** because they are required for transcription.

Once inside the cell, these messengers bind other molecules that react to the message. For example, estrogens enter the nucleus and bind to proteins there called *transcription factors*. The binding of the messenger changes the receptor's folding so that it can interact with DNA. Because estrogen's message often involves stimulating growth, the interaction with DNA may trigger a growth response, as in the uterine lining, which proliferates (grows) during the first part of the menstrual cycle to prepare for the potential arrival of a fertilized egg.

Second Messaging: Signals from Outside

While first messenger delivery inside the cell seems pretty straightforward, many first messengers cannot enter the contact cell. They must stay on the cellular doorstep and talk through the door, as it were.

Their conduit across the doorway—the plasma membrane—is a membrane protein that sticks out on both sides of the phospholipid bilayer. Because these proteins receive the messenger's signal while embedded in the membrane, we call them plasma membrane or cellular membrane receptors.

Bio Basics

Do you think a first messenger that can't enter the cell is hydrophobic or hydrophilic? If you said hydrophilic, you'd likely be right. That phospholipid bilayer is doing its job when it doesn't let large, water-soluble (hydrophilic) molecules in but instead leaves them on the doorstep to communicate through a protein keyhole.

In the case of cell signaling, when the first messenger binds its specific receptor protein poking out of the cell membrane, the interaction causes the embedded protein to change its folded conformation. That change happens on the inner-cell side of the membrane.

On the inside, that protein is in contact with other proteins. As it changes its conformation, it triggers the proteins next to it on the inside to change conformation, too. With this chain reaction, a signal crosses the cell's threshold without the messenger ever setting foot in the doorway.

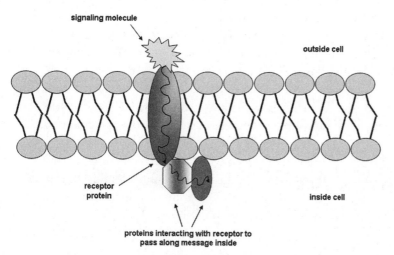

The messenger delivers the message to a membrane protein, which then transmits it inside the cell by changing shape and interacting with proteins on the inner membrane.

First messengers that get left on the doorstep include growth factors. These molecules are usually hormones made of amino acids, called peptide hormones. They're too large and polar to slip into the cell the way a lipid-based steroid hormone can. So they use those receptor proteins literally to get their message across. Most signaling molecules require specific proteins to receive their messages; not just any protein can do it.

How Is the Message Transferred?

We've talked before about how eukaryotic cells can be quite large. How is it possible for a first message to transfer its chemical instructions across the cell membrane and all the way to an inner structure like the nucleus? The answer is, a signaling cascade. This process of signal transfer, called signal transduction, involves a series of conversations along a chain of proteins that passes the signal from one to another until it gets where it's going.

A Cascade of Signals

Often, this series of signals, or signaling cascade, requires a second messenger inside the cell that can take up the instructions once they've crossed the membrane. The best-known example of such a messenger is cyclic AMP, or cAMP. Can you tell from its name that this is the "mono" version of two molecules we've already met, ADP and ATP? Yes, it's adenosine monophosphate, here acting as a messenger molecule.

One cAMP signaling pathway is the epinephrine pathway. The main steps involve the binding of epinephrine (also known as adrenaline) to its receptor protein poking out of the cell membrane. The adrenaline message is for the cell to break up some of its stored polysaccharides into individual glucose molecules. If a bear is chasing you, you want to get at that glucose for some ATP to power your suddenly needy muscles, and epinephrine helps you do that. But epinephrine cannot enter the cell, so it acts as the first messenger. The protein it binds to is from a class called the G-protein-coupled receptors.

Bio Bits

Our adrenal glands produce adrenaline, now called epinephrine. Its best-known role is in "fight-or-flight" responses, when its message is to shut down anything that's not mission critical for fighting or fleeing, like digestion. Epinephrine also acts as a neurotransmitter, passing signals through the gaps between nerve cells.

This protein gets its name because it sits right next to a G protein. A G protein gets its name because it's bound to a molecule of GDP or GTP. When epinephrine binds to the receptor, the receptor's conformation change triggers a change in an adjacent G protein. The G protein then releases the GDP in its active site and grabs a GTP. GDP is guanosine diphosphate, and GTP is the more unstable—and thus more energy-packed—guanosine triphosphate.

The GTP-bound G protein can now activate an enzyme next to it, adenylate cyclase (also known as adenylyl cyclase). This enzyme takes the old familiar ATP molecules and rips off phosphates to make the second messenger, cAMP. cAMP then interacts with the proteins in the signaling cascade that snip off individual glucose molecules from the cell's stores.

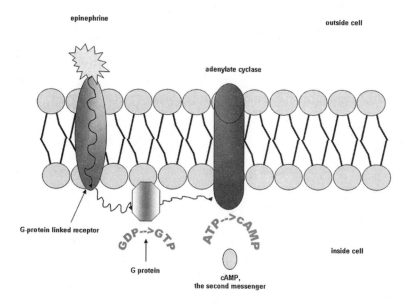

Epinephrine binds outside the cell. The receptor interacts with a G protein, which changes shape and exchanges a GDP for a high-energy GTP. Through a few more protein intermediates, the G protein activates adenylate cyclase, which metabolizes ATP to cAMP, the second messenger.

Think now about the process of the cell's response to epinephrine. It starts with one signaling molecule, the epinephrine. That molecule talks to one protein in the membrane, the G-coupled membrane protein. That single protein activates about 100 G proteins, which in turn activate about 100 adenylate cyclases.

Already, we've amplified our original single signal 100-fold. Each adenylate cyclase can produce two cAMPs, and then just about every step of the pathway after that increases the number of molecules that are activated. Finally, by the time the pathway reaches glycogen to break off glucose molecules, that single epinephrine signal will yield the cell about 100,000,000 glucoses for energy harvest. The signal transduction pathway makes quite a megaphone.

Bio Basics _____

The only second-messenger molecule we've discussed is cAMP, but other small molecules also act as second messengers, including calcium ions. The rules for a second messenger are (1) it is not a protein and (2) it is water soluble so it can diffuse quickly through the cell's cytosol and spread its message.

The Role of Phosphate

While the cAMP pathway is one way for a first messenger to get its signal deep into the cell, a few other kinds of pathways don't necessarily require an intracellular messenger service. Most of them involve moving phosphates around to change the conformation of proteins in the signaling chain. A phosphorylating chain reaction is a common way for a cell to pass the message along once it crosses the membrane.

This process can involve more adding of phosphates onto molecules to give them the right shape to interact with the next molecule in the series. Because the pathway consists of one phosphorylation event after another, we call it a phosphorylation cascade.

Ultimately, when all those phosphates get put on, something has to take them off. That something is a class of enzyme called phosphatases. These proteins help regulate the pathway by making sure it's not set to "on" all the time.

One final method of passing along a signal involves ion channel receptors. The receptor proteins receive their message from a signaling molecule such as a neurotransmitter, chemicals involved in nerve signaling. When they bind these ion channel receptors, the receptor changes its shape to create a channel. The channel usually allows only specific ions to flow in or out, like sodium, calcium, or potassium. When the _ligand_ releases from the receptor, the channel closes again.

def•i•ni•tion _____

A **ligand** is a molecule that interacts with the binding site of a protein, usually for signaling. Examples of ligands we've already encountered are hormones and neurotransmitters.

All good signals must come to an end, and the cell will reverse all of that binding eventually, leaving the proteins to return to their inactive form. Just as with every other step in these pathways, these "undo" steps have specialized enzymes of their own, such as phosphatases, to make sure the signaling gets turned off appropriately.

What Is the Response?

After much phosphate shifting, the signal finally arrives at its target destination. What happens next?

In an organism as complex as we are, cells are constantly signaling to one another. They do it to maintain the status quo, or homeostasis. They do it to defend against invaders. They do it to make our heart beat, our nerves transmit signals, our muscles contract, our bones grow, and our skin replace itself. Our cells signal when it's time to make new eggs or sperm, when it's time to make more thyroid hormone, and, for pregnant women near their due dates, when it's time to go into labor. We could fill an entire series of *Complete Idiot's Guides* with all the processes that involve cell signaling. Rather than doing that, let's just take a couple of good examples.

Growth

Previously you read a little bit about growth factors, the peptide hormones that signal from outside the cell as first messengers. They usually bind to receptor proteins poking out of the cell membrane and send their message.

You now know the drill: the protein changes shape, triggering something in the protein next to it inside the cell, which triggers more responses in adjacent proteins, and so on. Ultimately, this message reaches proteins in the nucleus. Their response is to kick start the process of cell division, called mitosis.

In Chapter 10, you learn about mitosis in detail, but here I can tell you two important things: (1) When biologists talk about cell proliferation, they mean cells dividing, or undergoing mitosis. (2) When an organism grows, that growth is usually achieved via mitosis. Thus, the response to the message from a growth factor is expected to be—mitosis!

Movement and Change

There's a time in your life that some biologists describe as the most critical moment of your existence. For several cell divisions after a sperm meets an egg, the cells are all essentially the same and haven't started to become different from one another. But at one critical moment, they do. One reason the cells "know" that it's time to become different is their communication with each other. They learn through messages from other cells exactly where they are in the bundle of cells and what they need to do next.

These signals can be paracrine—the diffusion of signaling molecules from one cell to a nearby cell—or they can result from direct interaction between cells. In some cases, cells must move elsewhere before they start becoming different. These cells get the message that it's time to let go.

What do they let go of? In many cases, they break down the glycoproteins they've been using to latch onto other cells. Once released, these cells are free to travel to their next destination. Often, they "know" where to travel because they're following chemical signals along a gradient in their surroundings, like cellular bloodhounds on a scent.

Once the cells get to their destination, they can receive messages from nearby or far away about what to do next. Let's say they've landed in an area destined to become the digestive system. Based on signals from surrounding cells and the rest of the organism, they'll focus only on producing the proteins that will make them digestive system cells.

Cell Death: A Special Case

The entire process of animal development is fascinating, and we'll cover it in more detail in Chapter 24. But there's another kind of cellular response to cell communication to talk about here, and it plays a prominent role in animal development, among other things. That process is called apoptosis (*apo* = separate from; *ptosis* = falling down), also known as programmed cell death or cell suicide.

Cells commit suicide? Yes, they do. Sometimes, they get the message from within. If the DNA suffers extensive irreparable damage, for example, the nucleus will send out self-destruct signals to the cell, and the process of apoptosis will begin. Another reason a cell might voluntarily kill itself is if protein folding has gone horribly astray; in this case, the endoplasmic reticulum, where proteins are assembled, hits the self-destruct button.

And then on occasion, the cell gets orders from another cell. It may be that the situation has become critically overcrowded and some cells must fall on the proverbial sword for the sake of the group. In that case, a molecule that signals "time to die" is sent to the chosen cell, kicking off the process of suicide.

Where does animal development come into all of this? Look at your hands. For a period of time in your embryonic development, you had webbing between your fingers. Some people retain some remnants of this webbing, but for most people, it disappears before birth. Except that it doesn't really "disappear." The cells that make up that webbing get a signal at just the right time in development that it's time to die, and they kick off the suicide cascade.

A cell achieves its own death by doing a couple of critically destructive things in response to the death signal. One is chopping up its DNA into bits and doing the same to its organelles. Essentially, it destroys its own factory components, meaning that all systems must shut down.

Then the cell shrinks around these bits, giving it the characteristic mottled appearance of a dying cell. A healthy cell has a nice round look to it. An apoptotic cell looks rather like a bundle of different size dark cotton balls. Those "cotton balls" are really cell bits that will be packaged into vesicles for destruction by other cells acting as the cleanup crew. Knowing as you now do how much nature likes to recycle and reuse, you can imagine the fate of the molecular components of the now-dead cell.

The Least You Need to Know

- Some cells communicate through direct contact, using gap or tight junctions, plasmodesmata, or glycoprotein interactions.

- Cell communication is tightly regulated based on specific recognition between the messenger and the recipient and involves receiving, transducing, and responding appropriately to the message.

- Some messengers, like steroid hormones, can enter the cell, delivering their message to the target molecules.

- Other messengers, like peptide hormones, are stuck outside the cell and must communicate across the membrane through second-messenger pathways.

- The pathways that transmit these outside messages to the inside involve intracellular second messengers like cAMP and often use phosphorylation cascades.

- The responses to the messages vary but can include growth, reproduction, and apoptosis, or cell suicide.

Chapter 10

Making More Cells

In This Chapter

- ◆ Producing identical cells
- ◆ The phases of mitosis
- ◆ The cell cycle
- ◆ Mitosis for growth, replacement, and reproduction
- ◆ Mitosis and cancer

For life to continue, cells must divide and make new cells. This chapter addresses the cellular copying machine of mitosis, which has the primary goal of producing identical copies of the original cell. Mitosis serves many purposes, from replacement and growth in single and multicellular species to reproduction for some organisms. Given the ultimate goal of exact copies, mitosis and cell division exist under tight regulation. If the regulatory machinery fails, however, the result can be disease, such as cancer.

Mitosis: The Cellular Copy Machine

Think about building a brick wall. You want bricks to do that, and it's your job to make them. Will you hope to make bricks that fit together neatly and uniformly? If you want your wall to look good, that's going to be your goal. To achieve it, a brickmaker will go to some lengths to ensure that the brick-making machinery turns out a series of identical bricks, each one a copy of the other.

Now think about building your skin's outer layer. It consists of cells. Do you want these cells all to be the same kind of cell, all doing the same thing? Unless you're interested in having a bit of liver growing here and there, or maybe some spleen, the answer is, yes.

Your body achieves this goal of sameness using a process of cell reproduction called mitosis, which means "division of the nucleus." The overall goal of mitosis is to create two cells that are identical copies of the original.

The Goal of Mitosis: No Change, Please

An organism grows by making more cells. Each of these cells is a copy of the very first cell that started the organism. The cells may eventually form different tissues, but in each cell, the DNA is exactly the same. Or, at least, it should be.

Mitosis involves a series of steps designed specifically to prevent any changes or differences between the original and the results. While this ultimately sounds like copying with a copy machine, the process is really more like typing a second copy of a document, using the original as a guide.

As with any typing job, the typist makes errors. Proofreading can catch some of these errors and fix them. However, if the typist—the cell—misses a mistake, that mistake becomes a part of the DNA for all the cells that follow. It may or may not be significant enough to change the nature of the cell. In some cases, the mistake is significant enough to cause diseases.

Given that a single one of your cells contains billions of letters—DNA bases—to type into a new copy, there's abundant opportunity for error. As commonplace as mitosis is—in you alone, you've made hundreds of trillions of cells using it—the real miracle is that it ever goes off without a hitch. Its sheer complexity makes its repeated success seem impossible. Yet, it does work, and here we all are.

Achieving the Goals: Growing Pains

Mitosis helped you grow into the complex, multicellular organism you are. And it does the work of replacement when your old cells die (apoptosis, anyone?). You turn over (replace) cells in your skin and bones, for example, at rapid rates. In fact, most of the skin you've got now is several generations from the skin you were born with.

We replace cells in many tissues, but researchers for many years thought that some tissues didn't replace, including nerve and heart cells. Recent findings show that nerve cells do regenerate under certain conditions and that in a lifetime, a person will replace about half of the heart cells they're born with.

Finally, some organisms reproduce using mitosis. A good example is yeast, a single-celled fungal eukaryote. Although it can engage in sexual reproduction (meaning shuffling of the genes of two organisms to create a unique individual), it also will use mitosis. The outcome, of course, is that the yeast essentially clones itself. People argue about the ethics of cloning organisms, but nature already does it with many species.

The most important step in achieving this perfect replica is to make a perfect copy of the DNA. And much of this process in eukaryotes relies on proper folding and packaging of the DNA. For some steps, the DNA must be carefully protected, especially during the crucial moments that the nuclear vault disintegrates. In other parts of the process, the DNA must be exposed so that the cell can copy it.

Although most organisms with backbones use sexual reproduction, there are exceptions. One is the desert grassland whiptail lizard, *Aspidoscelis uniparens*, an all-female species. They reproduce by mitosis—no sex involved—in a process called parthenogenesis, which means "virgin birth." To trigger egg formation and development, they engage in faux sex behavior.

DNA Packaging

We've learned that DNA is a long string of nucleotides (sugar + phosphate + base) attached to each other by covalent bonds. DNA usually occurs in double-stranded form, with two strings joined together by hydrogen bonds. The result is a kind of ladder twisted into a double helix, with the bases as the inner rungs, and the sugar-phosphate backbones along the sides. But the DNA in your cells isn't randomly wadded-up. It's carefully wrapped around proteins and packaged for safekeeping, especially during the dangerous process called mitosis.

All the DNA in your nucleus is your nuclear genome, or complete set of DNA. If we unfolded that DNA, it would be well over 6 feet long. Yet it fits into a nucleus in a cell that we can't even see with the naked eye. The careful winding of this long molecule around protein spools is the first step in the packaging of DNA in the cell. The DNA + protein combination is called chromatin because it stains colors we can visualize under certain conditions (*chrom* = color). The spools of protein that the DNA wraps around are histones. The table gives the different levels of DNA packaging, from the double helix to chromosomes.

Bio Basics

A genome is a complete set of DNA. It can refer to all of the DNA in an organism, as in the chromosomal DNA and extra-chromosomal DNA of bacteria, or to the DNA of specific organelles in an organism, such as the nuclear, mitochondrial, or chloroplast genomic DNA in eukaryotes.

Levels of DNA Packaging

Level	Features
Double helix	Nucleotide string
Histones	DNA wrapped around histone protein spools
Nucleosome	Histone-packed DNA linked to other spooled DNA
30-nm fiber	Nucleosome "beads" stacked into groups
Looped domains	30-nm fibers packaged into loops
Chromosome	Looped domains packaged and coiled together

For the cell to reach the DNA to type up a copy, this packaging must loosen up a bit. When the DNA is loosened up for copying, we call the available part of it euchromatin (*eu* = true). Some DNA doesn't loosen up even for copying, and we call that uptight DNA heterochromatin (*hetero* = different). But once the copying, or *replication*, process is complete, all of the DNA is packaged up tight again in preparation for the scary exposures of mitosis.

def•i•ni•tion

The process of copying DNA into DNA is called **replication**.

Stages: Follow the DNA!

Mitosis occurs in a set of stages that we've defined mostly for our benefit as observers. First, let's talk terms so we can follow the DNA around the cell as it is copied, joins a complex cellular square dance, and finds itself tugged literally to opposite ends of the cell.

DNA just sitting in the nucleus is just that, a molecule of double-stranded DNA packaged into a chromosome. After it's copied, though, each strand now has a partner attached to it. That partner is called a chromatid. The original molecule and its partner are together called sister chromatids, and they are (should be!) identical in sequence to each other.

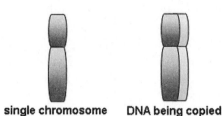

single chromosome   DNA being copied   sister chromatids

The chromosome on the left undergoes DNA replication (copying) to produce an identical chromosome joined to it at the "waist," or centromere (center). The two identical, joined chromosomes are now sister chromatids (right).

Sister chromatids hug each other via proteins called cohesins. Their closest region of attachment is the centromere, where the paired chromatids are also at their narrowest, thanks to the strong bonds there. Under a microscope, they have a tiny little waist where the centromere is. The arms of each chromatid extend above and below this centromeric area.

Parked at the centromere is also a special collection of proteins that play an important role in the mitotic DNA square dance. This structure is called the kinetochore. Each pair of sister chromatids has two kinetochores facing in opposite directions.

A Table of Confusing Mitotic Terms

Term	Definition
Centromere	Closest region of attachment between sister chromatids
Cohesin	Proteins holding sister chromatids together
Chromatid	A chromosome copy; pairs are "sisters"
Chromatin	Chromosomal DNA packaged with its proteins
Chromosome	Packet of DNA with proteins; highest level of DNA organization
Kinetochore	Collection of proteins at centromere that play a role in mitosis

Mitosis kicks off with the formation of a bunch of microtubules that behave like ropes with suction cups on the end, flinging out and sticking to kinetochores and then pulling them to the ends of the cells. In you, this assembly of microtubules, called the mitotic spindle, begins in organelles called centrosomes. There are two centrosomes, which start in the middle of the cell but then make tracks for the poles, tossing out the microtubules of the mitotic spindle as they go, like cowboys tossing lariats at a rodeo. An array of shorter microtubules they also spit out is called an aster (*aster* = star) because of its starburstlike appearance at each pole of the cell.

Now let's go through the steps of the mitotic square dance. First, the DNA tightens up. (It's already been copied, so these are paired sister chromatids.) Then the nucleus disintegrates, exposing the precious genetic material to the cruel world outside. The kinetochores on the sister chromatids become available for lassoing by the microtubules of the mitotic spindle. There is a major tug-of-war until all of the sister chromatids swing their partners, do-si-do, and line up single file along the cell's equator. They're preparing for the final steps of their dance here.

Finally, the lassoing microtubules start pulling the chromatids they've captured to opposite ends of the cell. The sisters separate, each copy moving to opposite poles. They are now officially individual chromosomes. Once each chromosome set has neared its polar destination, a nucleus forms around them at each end of the old cell. The old cell now has two nice nuclei packed with identical sets of chromosomes.

These steps occur in arbitrarily named phases, given in the table with their major events.

The Main Events of Mitosis, by Phase

Phase	Events
Prophase	Tightened chromatin, sister chromatids evident, joined at centromere. Mitotic spindle forms from centrosomes, starting at nucleus and moving toward poles.
Prometaphase	Nuclear envelope breaks up; microtubules latch onto kinetochores on sister chromatids.
Metaphase	Microtubules from each pole arrange sister chromatids on equator (metaphase plate), with centromeres aligned along plate, kinetochores of each chromatid pair oriented toward opposite poles.
Anaphase	Shortest phase. Cohesins sliced apart. Microtubules tug sister chromatids apart to opposite poles as cell elongates. When anaphase ends, each pole will have a complete chromosome set for that organism.

Phase	Events
Telophase	Two new nuclear envelopes quickly form, one around each set of chromosomes at the poles. The cell now has two nuclei! Mitosis is complete.

The Cell Cycle: Mitosis Is Just a Chapter

What? you may be thinking. We just walked our way through making a cell with two nuclei. Aren't we supposed to get two new cells out of this? Yes, we are. What I've kept secret from you this entire time is that mitosis is really just a small, short, but very important part of a larger process called the cell cycle. The division of the nuclear material into two new nuclei is obviously critical. But mitosis isn't the end of it.

Biohazard!

Don't let the feminine terminology mislead you. Biologists refer to mother cells and daughter cells and sister chromatids. These terms simply express relationships and have nothing to do with the sex of the organism. Males have mother and daughter cells and sister chromatids, just as females, or organisms that are both sexes or neither sex, do.

Usually, following mitosis, the final step of cell division takes place. This process is called cytokinesis (*kinesis* = splitting, so cytoplasmic splitting). Once the nuclear envelopes are safely reformed around the precious DNA contents, we can turn our focus back to the cell midline. There, a drawstring effect is taking place, with actin microfilaments serving as the strings. They draw tighter and tighter around the midline of the mother cell until the cytoplasm pinches off completely to yield two daughter cells, each with a nucleus and full DNA complement.

Interphase: G1, S, G2

But that's not all. Mitosis and cytokinesis together still make up only a very small part of the cell cycle. Nature, like any good project manager, knows that the prep work requires the most time, and we find that most of the time the cell cycle requires is devoted to this prep work.

Together, mitosis and cytokinesis form the "M" phase of the cell cycle. We've just learned what happens in that phase. But there's a much longer prep phase called interphase.

Interphase occurs in three stages. The first is G1. (There's that scientific creativity again.) This gap phase (I really like to think of it as a growth preparation phase) involves prep work for the copying process. The cell makes the necessary proteins and gets the copying machinery ready.

The S phase is the synthesis phase, when the DNA copying takes place. From this phase on, we have duplicate chromosomes as sister chromatids. Thus, once a cell enters into G1, it usually goes on to complete the entire cycle, or it would just be wasting energy. We call the entry into G1 a checkpoint because that's the time when the decision to cycle or not to cycle gets made.

After S phase comes the second gap phase, G2. In this phase, organelles like mitochondria double in number in preparation for the upcoming production of two daughter cells.

The entire process of interphase requires about 95 percent of the time necessary to complete a cell cycle. Nature, as you can see, is a smart project manager. Get all the ducks in a row, all the equipment and materials in place, devote your time to the prep work, and then watch the crucial phase—M phase—run like clockwork. Most of the time.

Cytokinesis Optional

You may be wondering why we make a distinction between the mitosis and cytokinesis parts of M phase. Why not just call it all mitosis and make it easier on poor students like you? Well, not all cells that undergo mitosis undergo cytokinesis.

A great example is you. The muscle cells of your body often have more than one nucleus, or are multinucleated. They create their multinucleated condition by mitosing but avoiding cytokinesis. This setup works for muscle cells by allowing rapid signaling. Sometimes, however, other kinds of cells end up with more than one nucleus because of a disease process, such as cancer, that has disrupted cell cycling.

Rest (G0)

Remember that G1 checkpoint I mentioned a moment ago? The cell has a choice at this checkpoint—to cycle or not to cycle. Some cells cycle a lot, such as the cells that produce your skin or your immune cells. But other cells rarely cycle, if at all. They are the cells that made that choice at G1 to stay just as they are, thank you very much.

When a cell makes a choice not to cycle at G1, it slips into a stage of rest, called G0. Most of our cells are in this stage, because they're not required to reproduce. We used to think, for example, that all nerve cells existed permanently in this condition, although we now know that they sometimes do divide. Cells may be awakened from G0 if the organism requires some new cells from it. And when a cell decides to just wake up on its own and become a rogue cycler, the result is cancer.

Binary Fission: Bacteria and Mitochondria

When the yeast I mentioned above divide using mitosis, they are undergoing a process we generally refer to as binary fission. Fission means splitting, and binary means into two, so this process is splitting something into two. But more specifically, we tend to talk about binary fission occurring in prokaryotic organisms using a process that is not mitosis. Because organelles like mitochondria were themselves once prokaryotes, they also undergo binary fission when the cell is cycling. It's how we double the number of mitochondria for equal distribution into the two new daughter cells.

Binary fission in bacteria involves making a copy of its circular chromosome while the cell gets longer in preparation for splitting into two cells. As the copying completes, the elongating cell starts to narrow at the middle. At the same time, the bacteria produce the components to build a new cell wall. When the pinching off at the middle is complete, the result is two new bacterial cells, each with a complete circular DNA copy and a nice new cell wall.

The Cell Cycle and Cancer

Cancer is best described as mitosis gone out of control. There's that crucial G1 checkpoint when the cell makes the decision to go forward or go rest. If the cell makes a "bad" decision, the wrong decision to go forward, then there is growth when there shouldn't be. When the growth has no limits, that's cancer.

It may not surprise you to learn that kinases are involved in this decision. These kinases hang out in the cell, waiting for someone to signal them to wake up and kick off some cycling. The cycling signal comes from proteins called cyclins. Because the kinases rely on cyclins as a sort of regulatory alarm clock, the kinases are called cyclin-dependent kinases, or Cdks. If they're active, the signal at G1 is "go."

A protein that is central to checking the cell cycle is p53. This molecule regulates production of another protein called p21. p21 keeps Cdks sleepy and inactive, and the signal at G1 is "rest." If p53 doesn't work right, however, then p21 doesn't get made.

The Cdks wake up and become active when they should still be asleep. The cell divides out of control, passing its mutant p53 gene to its progeny.

Bio Bits

The best-known cell cycle protein is probably the humbly named p53. Had its discoverers foreseen its ultimate significance as the "guardian of the genome," they might have given it a livelier handle. And even that name is inaccurate. The P correctly indicates that it's a protein. But the 53 refers to what researchers initially identified as its mass (53 kilodaltons in molecular mass). As it turns out, its mass is closer to 43 kilodaltons.

We call p53 a tumor suppressor gene because when it's behaving normally, it suppresses tumor formation by regulating cell cycling. The p53 protein actually serves several protective purposes in cell cycling, including signaling when DNA damage is so severe that the cell should commit suicide.

Cancer comes in stages. If the dividing mass of cells stays in one place and doesn't invade nearby tissue, it often can be surgically removed. The stage of severity increases if the tumor cells invade nearby tissues, and such tumors are usually more difficult to remove completely. If some of the cells break away, spread to other parts of the body, and divide there, they have metastasized. Metastasis is often the worst stage of any cancer and the hardest to manage successfully.

The Least You Need to Know

- The goal of mitosis is to produce identical daughter cells for growth, replacement, and reproduction.

- The primary events of mitosis are divided into prophase, prometaphase, metaphase, anaphase, and telophase.

- These mitotic events mark the movement of chromosomes/sister chromatids lining up along the cell equator and separating into equal sets.

- The cell cycle also includes the much longer interphase, divided into three stages: G1, S, and G2.

- Prokaryotic organisms use binary fission to divide.

- Errors in proteins that regulate the cell cycle can result in out-of-control growth, which can manifest as cancer.

Part 3

Genetics

Sexual reproduction and genetics go hand in hand because the latter led us to the former. Mendel's work with pea plants uncovered several inferences about how sexually reproducing species transfer hereditary material to off-spring, inferences that found later confirmation in actual observations of sexual cell division. And then, of course, identifying DNA as the molecule of heredity has opened the door onto a world of discovery, including how nature takes a collection of four different kinds of molecules to produce all of the diversity of life we see around us.

Chapter 11

Making More Cells:
Special Edition

In This Chapter

- The goals of meiosis
- The events of meiosis I
- The events of meiosis II
- Meiosis in males vs. females
- Fertilization in brief

In Chapter 10, we learned about mitosis—how cells closely regulate chromosomal copying and movement to produce two identical cells. Well, now it's time to throw all of that out the window because we're moving on to meiosis. Where mitosis was all about perfect copies, meiosis is all about mixing things up and cutting things in half. Where mitosis was about making two cells with two perfectly copied sets of chromosomes, meiosis means making four cells, each with a single mixed-up chromosome set. Why all the mixing up? If you look around you, you'll see the outcome of all that shuffling: the differences between you and me and everyone else.

Meiosis: A Very Special Process

The goals of meiosis are practically the opposite of mitosis: DNA shall be shuffled in meiosis. No chromosome will be the same as the parent cell chromosomes. The new cells that result from meiosis, a set of four cells we call gametes, won't even have the same number of chromosomes as the parent cell. Nope. They'll have half the parent number. There's a good reason for that, as you will see.

Shuffling DNA. Cutting chromosome number in half. Welcome to the strange and wonderful world of meiosis and sexual reproduction.

The Goals: Variation and Reduction

You may be asking yourself, "What is the point of shuffling DNA?" Evolution and natural selection work on the differences among organisms to select those with traits best adapted for a specific environment. Without gene shuffling, nature would have little variation to choose from. Without gene shuffling, species might not have individuals that could adapt to a changing environment, and adverse environmental changes would simply wipe out life.

If all organisms relied on mitosis to reproduce, most organisms would be clones. Sexual reproduction, meiosis, and the shuffling with these processes yield much of the diversity you see among members of the same species and between species. Variation is not just the spice of life—it's a necessity for much of life to continue.

And reduction of the species chromosome number is required for sexual reproduction to work out right. The final step of sexual reproduction is fertilization, when a cell from one individual fuses with the cell of another individual. Simple math will tell you why the chromosome number needs to be reduced.

> **Biohazard!**
>
> Don't make the mistake of thinking that all clones are bad or that variation doesn't happen with asexual reproduction. Plenty of asexually reproducing organisms like bacteria have been successfully cloning since life on Earth began. Through mutation and some interesting methods of genetic transfer, they've also accumulated their variations.

Take people, for example. Our cells each have 46 chromosomes. If we used cells with the full chromosome number and fused two together, we'd make a cell with 92 chromosomes. Even 1.5 times the number of chromosomes is lethal during fetal development in humans.

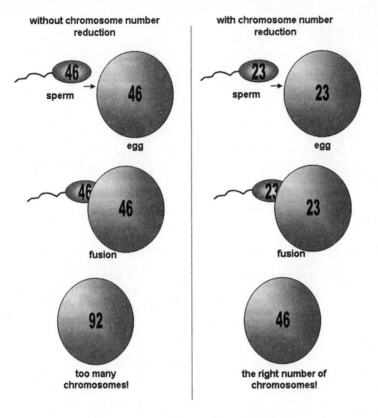

without chromosome number reduction

with chromosome number reduction

sperm → egg

sperm → egg

fusion

fusion

too many chromosomes!

the right number of chromosomes!

Meiosis achieves reduction in chromosome number, essential for sexual reproduction involving the fusion of two cells.

Meiosis involves two cell divisions. The final four cells that result each end up with half the number of chromosomes of a regular body cell. Thus, when a person's *germ cells* divide by meiosis, the result is four (well, maybe four—more on that in a minute) cells, each with 23 chromosomes. That way, when one of these cells fuses with a meiotic daughter cell from another person, the resulting cell will have the magic number of human chromosomes: 46. And it will have a unique DNA complement, the result of DNA shuffling during meiosis in each parent and the new combination of parental chromosomes.

def•i•ni•tion

A **germ cell** is the cell that undergoes meiosis to produce gametes, the products of meiosis. Gametes are called sperm in males and eggs in females.

Achieving the Goals: DNA Shuffling

The cell achieves DNA shuffling during three different steps of meiosis. But first, let's talk terms. You may recall from Chapter 10 that chromosomes are tidy packets of long strands of DNA. When the cell copies DNA, each chromosome finds itself attached to its identical copy; we call that pair of chromosomes sister chromatids. So far, so good.

Here's where things get fancy. You have two basic cell types in your body. The cells of your organs and tissues are somatic, or body (*soma* = body), cells, born by mitosis, dividing by mitosis. But most people also harbor a special group of cells, the germ cells. These cells are born by mitosis but divide by meiosis.

The end result of meiosis is four cells, each with half the species chromosome number. How does the cell do it? Once again, we turn to simple math.

Meiosis occurs in two parts, creatively named meiosis I and meiosis II. The main event of meiosis I is the separation of pairs of chromosomes. This is where things get really confusing. The pairs that separate from each other in meiosis I are *not* sister chromatids. Nope. Those stay stuck to each other until meiosis II.

So what separates? Well, let's back up to before you were conceived. A sperm swam toward an egg (or was injected into one if it was *in vitro* fertilization). The sperm carried a chromosome complement of 23 chromosomes. Those were the paternal chromosomes you got from your biological father. The target egg also carried a complement of 23 chromosomes. Those are the maternal chromosomes you received from your biological mother.

Bio Bits

Like all mammals, we receive a chromosome complement from each biological parent, one maternal, one paternal. But is it possible to bring together two maternal complements and have normal development? Most research suggests no. Parental sets complement each other during development, and one requires the other. However, one recent group induced normal-appearing development from a mouse cell containing two maternal chromosome sets. If sexual reproduction were to no longer require a second sex, what would that mean for the future?

For each set of 23 chromosomes, there are 22 nonsex chromosomes we call autosomes, and then a sex chromosome (22 autosomes + 1 sex chromosome = 23 chromosomes). In the egg, the sex chromosome will be an X. In the sperm, it could be a Y or an X.

The 22 autosomes are numbered 1 through 22. Chromosome 1, for example, is a huge chromosome with a ton of genes on it. Chromosome 21, on the other hand, is a tiny chromosome with many fewer genes on it.

When the sperm fuses with the egg, its 23 chromosomes will have their same-numbered partners in the egg. Thus, the cell that results from the fusion now has a chromosome 1 pair, a chromosome 2 pair, and so on. Each pair consists of a paternal and a maternal chromosome. These pairs, one from mom and one from dad, are called homologous pairs, or homologues.

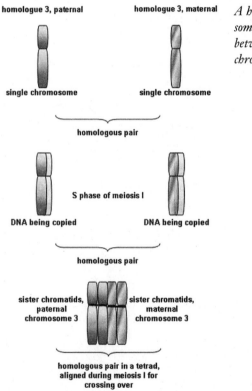

A homologous pair of chromosome 3 shows the difference between homologues and sister chromatids.

Before meiosis I, each chromosome will undergo DNA synthesis to produce a sister chromatid that stays attached to it throughout meiosis I. Thus, we have 46 chromosomes organized into 23 homologous pairs, and each of the 46 chromosomes has a sister chromatid attached to it. In meiosis I, the homologous pairs line up next to each other—two lines of 23 chromosomes each—and then separate into the two daughter cells, each still with its sister chromatid attached. The main event of meiosis II is the separation of these sister chromatids.

> **Bio Basics** _____
>
> You should especially remember a few meiotic highlights. First, genetic shuffling, which introduces variation, takes place in meiosis I during crossing over and when homologues randomly separate into daughter cells. Meiosis II does a bit more shuffling as chromatids separate into the final meiotic products. Finally, the major event of meiosis I is that homologues separate, while the major event of meiosis II is that sister chromatids separate.

Stages: Follow the Parental Genes!

Let's take each meiotic division one at a time. There are some big events that go down in meiosis I before the key event of homologue separation takes place.

First, we have DNA synthesis. Then the phases of meiosis I are similar to those of mitosis: prophase, metaphase, anaphase, and telophase. But there are some notably meiotic-specific events to highlight in a couple of these.

First, there's prophase I. Here, we see the first shuffling of the DNA (variation alert!). Let's take a sample scenario. Paternal chromosome 3 with its sister chromatid attached is hanging out in the nucleus. It's near maternal chromosome 3, which also has its sister chromatid attached.

If the paternal and maternal homologues align with each other, they can exchange chromosome bits, as the figure on the following page shows, in a process called crossing over. Thus, paternal 3 may switch out a chromatid tip with one from maternal 3. You should pause in amazement: already, before any division has even happened, the chromosomes that the offspring will inherit are now _already different from their parents' chromosomes._

The two homologues line up gene to gene in a process we call synapsis. Usually, the tips of the chromosomes, being the most available, are the most likely to cross over. The switch out during crossing over should be an even exchange of the same length of DNA covering the same section of gene for each member of the homologous pair. If you imagine the genes designated from the tip as Gene A, Gene B, and Gene C, then paternal 3 and maternal 3 might do an even exchange of each other's Gene A at their tips.

A pair of chromosomes that shouldn't and usually doesn't do crossing over is an X and a Y. They're not even technically a homologous pair since homologues are pairs of the _same_ chromosomes. These two normally do not align and cross over because they do not have the same genes to align.

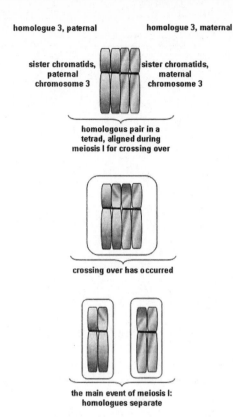

homologue 3, paternal   homologue 3, maternal

sister chromatids,   sister chromatids,
paternal   maternal
chromosome 3   chromosome 3

homologous pair in a
tetrad, aligned during
meiosis I for crossing over

crossing over has occurred

the main event of meiosis I:
homologues separate

In meiotic division, DNA synthesis happens once, in meiosis I. Then homologous chromosomes line up and exchange genetic material via crossing over. The main event of meiosis I is that the homologues then separate. Now imagine this graphic with 23 pairs of chromosomes, instead of one pair, and you'll imagine human meiosis.

Bio Basics

While crossing over is expected and generally useful in meiosis, it is not supposed to happen in mitosis. Homologues do not align in mitosis, so any crossing over will likely occur between nonhomologous chromosomes. The outcome is an unequal exchange of unrelated genes, which often ends in a disease state or failure of development.

In metaphase I, you can really visualize best the difference between homologues and chromatids. In metaphase of mitosis, each chromosome with its sister chromatid attached lines up in a single file of 46 chromosomes along the cell's equator. In meiosis I, what we see is *two* lines, each consisting of 23 chromosomes lined up side by side. In anaphase I, these homologues will be pulled to opposite ends of the cell, 23 to one end and 23 to the other end.

The fictitious organism depicted here has two sets of chromosomes for a total of 10 chromosomes. In mitosis, those 10 chromosomes will line up single file on the metaphase plate, each with a sister chromatid attached. In meiosis I, the homologous pairs line up in five pairs on the metaphase plate, each also with its sister chromatid attached.

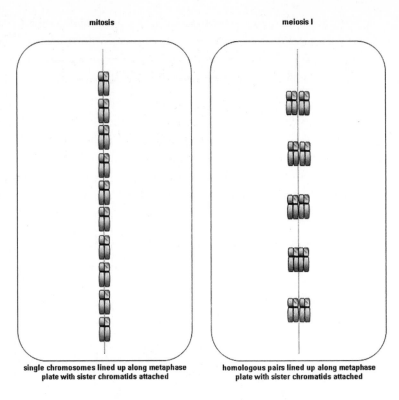

mitosis

single chromosomes lined up along metaphase
plate with sister chromatids attached

meiosis I

homologous pairs lined up along metaphase
plate with sister chromatids attached

And here's another major shuffling event (variation alert!): The cell doesn't pull 23 paternal homologues to one and 23 maternal homologues to the other. No. The cell mixes the homologues up. Thus, paternal homologues 1, 4, 6, 8, 9, 11, 12, 14, 20, 21, 22, and X might get pulled to one end of the cell along with maternal homologues 2, 3, 5, 7, 10, 13, 15, 16, 17, 18, 19, and another X. Count them. That's a full 23 homologues being pulled to one end of the cell, but it's a mix of paternal and maternal homologues. The other end of the cell gets the remaining mixed set of 23 homologues. We call this tendency of homologues to assort into daughter cells independently of parental origin *independent assortment*.

def•i•ni•tion

Independent assortment refers to the movement of homologues in meiosis. Even though one set of the pair is paternal in origin and the other set is maternal, when they separate during meiosis I, they do so independently of origin. The result is a mix of paternal and maternal chromosomes in the resulting cells.

Telophase I comes next, followed by cytokinesis. The final result of meiosis I is two daughter cells, each with a complement of 23 homologues. And each homologue still has its sister chromatid attached. That sets the stage for the major event of meiosis II: those attached-at-the-hip chromatids finally separate.

Meiosis II also has prophase, metaphase, anaphase, and telophase. This time, there is no DNA synthesis—that's been done. And in metaphase II, the chromosomes, each with its sister chromatid attached, line up in a single file of 23 chromosomes along the metaphase plate. Microtubules at each pole latch onto the chromatids at the kinetochore. A complete set of 23 chromatids—now officially chromosomes—is pulled to each end of the cell in anaphase II. This final step of chromatid tugging is also the third point of DNA shuffling in meiosis (variation alert!).

homologue 3, paternal   **homologue 3, maternal**

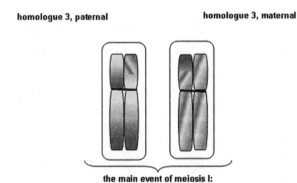

the main event of meiosis I:
homologues separate

This is the second cell division of meiosis, when the sister chromatids are pulled apart and assort into the four cell products, or gametes, of meiosis. Again, imagine this multiplied by 23 chromosomes, and you'll have imagined the outcome of human meiosis!

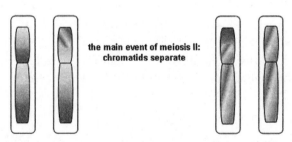

the main event of meiosis II:
chromatids separate

Four gametes each with a single shuffled chromosome set are the final outcome of meiosis.

Each daughter cell from meiosis I undergoes this process and divides, each producing two daughter cells, or gametes. Each gamete contains 23 chromosomes. Reduction of chromosome number is now complete, and we have a total of four cells (maybe—more on that later), each with a half-complement of chromosomes. Mission accomplished.

Mars vs. Venus: Meioses Compared

You may have noticed that I keep hedging a bit when I talk about the final product of meiosis. What's with this "four—maybe" business? Well, it's all about energy and who's got to put the most into the cell that results from fertilization. Among organisms like us, the "who" that does this job is the egg. For that reason, when a germ cell in a female undergoes meiosis, the result is one big fat healthy gamete and three toss-offs called *polar bodies*.

def•i•ni•tion

> A **polar body** is a product of female meiosis that contains a nucleus but very little cytoplasm. The cytoplasm has been allocated through asymmetrical cell division to the cell that is destined to become the egg. Female plants and animals both produce these polar bodies.

Males: Four Functional Gametes

A sperm is a pretty basic little cell. Its job is to get made by meiosis, emerge on cue, and then swim like crazy to find the egg. Once there, it's got to win the race to the prize and find its way through the egg's fairly major defenses. To achieve all this, the sperm needs its half-chromosome complement (23 in humans), a good energy source in the form of mitochondria, and a *flagellum* for the motor to swish around. In its tip, the sperm is equipped with a special packet of enzymes that, if it's the lucky winner, it will use to bore through the egg's defenses.

The sperm cell has just got the one job, so it doesn't have all of the bells and whistles of a full-blown, *diploid* somatic cell. Just the DNA, the motor, the tail, and a packet of enzymes for drilling a hole in the egg on contact. When a male animal makes sperm by meiosis, the result is four functional *haploid* cells, ready to swim for it when the right time arrives.

def•i•ni•tion

A **flagellum** is a string of proteins linked to motor proteins that move it around. It aids in locomotion in many cell types, including eukaryotes and bacteria, although the structure and motion differ among types. The flagellum of a sperm moves the cell through a back-and-forth whipping action.

The suffix *–ploid* refers to sets of chromosomes. Our cells are primarily **diploid,** or have two sets of 23 chromosomes each for a total of 46. But the gametes that result from meiosis are **haploid,** or have half the typical number of chromosome sets: one set of 23 chromosomes.

Females: One Egg and Three Polar Bodies

Those speedy little sperm cells are swimming for the grand prize in sexual reproduction: the egg. When you see pictures of sperm around an egg, you realize how much the enormous egg dwarfs the little motorized swimmers.

Maintaining that size requires a lot of energy input. When females make eggs, they don't bother producing four functioning gametes. The egg isn't in a race for a prize—it is the prize. No need to make millions of them at once to compete. So the energy goes into producing one good, healthy egg with meiosis.

But the process still involves two divisions that ultimately yield four products. The other three products are the polar bodies, which have given everything important they have cytoplasmically speaking to the big egg and will then fade away into oblivion. Then the big egg sits and waits for a bit—in humans, usually just a day or so—to see if any visitors show up.

Fertilization: Back to Our Regularly Scheduled Mitosis

We've briefly described the process of fertilization earlier, but we haven't talked about what fertilization achieves. When the sperm fuses with the egg, the nuclei of each gamete start to migrate to one another and fuse. The cell that results from this fusion is called a zygote. The minute it forms, this new cell, with its new and unique DNA complement, will start to divide. The instructions for dividing come from the abundant cytoplasm provided by the big fat egg, which gives the new cell everything it needs to get going.

And then, the zygote begins anew the process of cell division. Because it's now a growing organism, the process it uses is mitosis, and the goal is to make a series of identical cells.

The Least You Need to Know

- The purpose of meiosis is to shuffle genes and produce cells with half the chromosome number.

- Genetic shuffling occurs during crossing over and separation of homologues in meiosis I and of chromatids in meiosis II.

- The main event of meiosis I is homologues separating, and the main event of meiosis II is the chromatids separating.

- In meiosis, males typically produce four gametes (i.e., sperm), while females make a large egg and three polar bodies.

- At fertilization, the chromosome number is restored with the fusion of the nuclei of two gametes into a zygote.

Chapter **12**

Mendelian Genetics

In This Chapter

- ◆ Mendel and the pea plants
- ◆ Punnett squares
- ◆ Laws of segregation and independent assortment
- ◆ Probability rules
- ◆ Mendelian inheritance
- ◆ Chromosomal and other inheritance

Gregor Mendel used tens of thousands of pea plants to make several accurate observations about inheritance decades before scientists demonstrated any of it with modern techniques. With his plants in an abbey garden, he established two laws of inheritance that reflect directly the events of meiosis. His brilliance lay in his method and his math.

Math also underlies our calculations of inheritance today, determining probabilities about who will inherit which genes. The associations between genes and the characters they give us are more complex than what Mendel derived from his pea plants. They extend beyond the concept of a single gene to multiple genes and even inheritance linked to entire chromosomes or to mitochondrial DNA.

Mendel: Devoutly Statistical

Some students likely walk away from learning about Mendel with the impression that he was just lucky. But he was smart, he knew his plants, and he was a true scientist who interpreted the biological underpinnings of his statistical results.

He also was a monk who lived in a monastery for most of his adult life, where he worked as a teacher and cultivator of plants. The peas eventually led to his posthumous fame.

Mendel's Choice: Observation, Not Luck

Mendel didn't choose these plants by accident. He knew from experience that they had critical features that made them an excellent choice for his study organism. He could self-cross a plant, meaning that he could use a plant's *pollen* to fertilize the same plant's *ovule*. The pea plants had a set of binary traits, meaning that they were either green or yellow or had either wrinkled or smooth peas, nothing in between, no "blending." And they had short generation times, meaning that they matured fast and he'd get his results quickly.

def•i•ni•tion

In plants, **pollen** are sort of analogous to sperm in animals, and **ovules** are vaguely analogous to eggs.

Also, Mendel was a mathematician who could interpret his statistical data. He generated a lot of data, ultimately working with about 29,000 plants to achieve his grand opus on genetics. This opus, entitled "Experiments on Plant Hybridization," attracted almost no attention. It remained rarely cited and of little interest for decades until resurfacing in the twentieth century, when scientists began to appreciate its genius.

Why is it a work of genius? Because with nothing in the way of molecular biology tools, with no information about genes, Mendel used 29,000 pea plants to figure out how parents pass on their traits to offspring.

Now Mendel's work is considered so important that we've named an entire portion of the study of genetics after him. We call it Mendelian genetics, the genetics of heredity from parent to offspring. Mendel is now known as the father of genetics.

Peas: Traits, Phenotypes, Genotypes

In casual conversation, we might refer to a red-headed person from a family of red-heads as having that family's trait. For once, there's not too great of a divide between the common use of a biological term and the biological meaning of it: a trait is a feature of an organism, one that is usually largely *heritable*.

So red hair is a trait, as is having freckles, or hazel eyes. Blood type is another example of a trait that frequently makes an appearance in genetics textbooks—and yep, you're going to find it here, too. Traits often arise simply as the result of having specific genes. But sometimes, traits are more complicated and also subject to environmental influences.

def•i•ni•tion

Heritable designates a trait that can be passed genetically from parent to child.

Height is a good example of a complex trait: while much of your height can trace to how tall your parents are, diet has a great influence, too. Behavior is another good example of a complex—very complex—trait.

A trait can often manifest in different ways. Height is a continuous trait, meaning that there are infinite gradations of height, such as 5 feet, 2 inches, or 5 feet, 2.1 inches, and so forth. Some traits, however, are more straightforward or discrete. The peas, for example, could be either yellow or green. We call the different variants of a trait phenotypes.

Mendel's choice of binary traits helped keep his statistical data reasonably clean so that he could tease out the patterns of inheritance. What Mendel could only infer from his results was a plant's genotype. The genotype is the genetic combination that underlies the phenotype. Because of environmental influences, not everyone with the same genotype will have exactly the same phenotype, and quite often, there are several possible genotypes underlying a single phenotype.

 Bio Basics

A phenotype is, strictly speaking, any observable characteristic of an organism. Note that "observable," doesn't necessarily mean "with the eyes." A phenotype could be, for example, something detectable chemically, like metabolic by-products.

The Punnett: What the Gametes Carry

For our purposes, we're going to stick with a single gene pair for a given trait. That keeps things from getting too confusing. But what do we mean by "gene pair"? You may not realize it, but if you read Chapter 11, you've already traced the path of how a gene pair for a particular trait comes together. When the sperm fuses with the egg in fertilization, each gamete brings each member of a gene pair for a given trait. These gene pairs occur on pairs of homologous chromosomes.

Bio Bits
Who was this Punnett fella? His name was Reginald Punnett, and we owe much of our deep appreciation for Mendel's work to him and his colleague and co-geneticist, William Bateson. Punnett also is credited with writing the first genetics book and introducing genetics for the first time to a popular audience.

Mendel identified this pairing of a gene from each parent using his pea plants. Let's take a look at the trait of pea color. Mendel's choices were yellow peas or green peas. Those are our two possible phenotypes. Underlying each phenotype is a pair of genes. If the plant has even one copy of the gene variant for yellow in its pair, it will be yellow. If the plant has no gene variant for yellow in its pair, it will be green. Mendel crossed many green pea plants to figure out that they carried only the gene variants to be green. And he crossed many yellow pea plants to figure out that some of them carried a hidden ability to pop out a green pea offspring.

Pretty simple so far? Let's make it complicated. We call different forms of the same gene alleles (*allo* = different). In pea plants, the color trait has two possible alleles (Y for yellow and y for green).

Take a look at the Punnett square (a) in the figure on the following page. Now think back to meiosis. Let's say you have a male parent plant, Bob, carrying the alleles Y and y. When Bob produces pollen by meiosis (pollen carries the plant equivalent of sperm), what alleles can Bob's pollen carry? Remember that each pollen gets only one of the two possible choices.

If you said "Y" or "y," you were right! Parent plant Bob was Yy. When the homologues and then the chromatids separated in meiosis, each gamete got only one of those two alleles. In fact, two of the four gametes each got a Y, and two gametes each got a y.

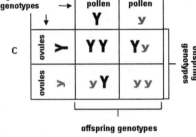

For the top Punnett (a), try to fill in the ? with the gamete genotypes that a male Yy plant and a female Yy plant would make. The middle Punnett (b) shows the answer. Now try to fill in the offspring genotypes that would result with each possible pairing of the gametes. The bottom square (c) shows the answers. The phenotype ratio of the offspring is 3 yellow to 1 green; the genotype ratio is 1 YY: 2 Yy: 1 yy.

Our female parent plant, Doris, produces ovules (the plant equivalent of an egg) by meiosis. She is also Yy. Her gametes also will be either Y or y.

Look at the Punnett square (a) again. The pollen Bob produces can carry either a Y or a y. The ovules that our Yy Doris can produce by meiosis also can carry either a Y or a y. We write these possibilities in the Punnett square as shown in (b).

Now look at the filled-in Punnett square (c). The pollen and ovule boxes show the possible *genotypes of Doris and Bob's gametes.* The other boxes show the *offspring genotypes* that would result if a given pollen from Bob met a given ovule from Doris. If pollen Y met ovule Y, the genotype of the resulting baby plant would be YY. If pollen y met ovule y, however, the genotype of the resulting baby plant would be yy.

In these plants, having even just one Y in the pair means the plant will make yellow peas. Three of the offspring genotypes have at least one Y in the gene pair. So if Bob and Doris reproduced together, three out of every four of their baby plants would be yellow. The lone green pea offspring of Bob and Doris is represented by the genotype of yy.

With this cross of a Yy Bob with a Yy Doris, their offspring will be 3 to 1 yellow to green, or have a 3:1 phenotype ratio. The genotype ratio of their offspring will be 1YY to 2Yy to 1yy, or 1:2:1. Mendel crossed hundreds of yellow pea plants and figured out that when they produced 75 percent yellow pea offspring and 25 percent green pea offspring, those parent plants must each have a green pea allele hidden away with their yellow pea allele.

Bio Basics _____

In genetics, we name generations to avoid confusion. The parental generation, the one we start with, is the P generation. Their offspring are the F1 (first filial) generation. If you cross F1 with another F1, you'll get F2 (second filial).

Law: Thou Shalt Separate from Thy Partner

When Bob makes his pollen through meiosis, he's working with genes that he himself inherited from each of his parents. Bob evidently inherited a Y from one parent (let's say it's Bob Sr.) and a y from his mother (Ethel). Thus, Bob's pair of genes, Yy, came from Bob Sr. and Ethel. In fact, they formed a homologous pair in Bob. And when Bob makes gametes, that homologous pair must separate during meiosis.

Through his plants, Mendel figured out that the parental gene pairs had to separate somehow into singles, with the gene pair then restored at fertilization. He had no way of knowing that this separation of the pair starts during meiosis I, when the homologues separate and finishes in meiosis II when the chromatids separate. But he knew enough to formulate a genetics law, the Law of Segregation, to describe exactly that process.

This law states that when an organism makes gametes, each member of its gene pairs will separate into different gametes. That's just what happened with Bob's Yy! His homologues separated—segregated—during meiosis I, and his chromatids separated—segregated—during meiosis II into different gametes.

Law: Thou Shan't Follow Other Genes Around

When Mendel studied his pea plants, he looked at several traits. (He didn't focus only on pea color.) And he noted that the phenotypes for different traits didn't travel in packs. For example, the green phenotype for pea color didn't always show up with the wrinkled phenotype for pea shape. Sometimes, green peas were smooth, or yellow peas were wrinkled.

Once again, without access to information about meiosis, Mendel figured out exactly what was going on. The genes for pea shape and the genes for pea color were not attached to each other and behaved separately from one another. They do this because they are on different chromosomes, and those chromosomes assort independently from each other in meiosis.

Although he didn't know all that chromosome stuff, he formulated from his observations the Law of Independent Assortment, in which the phenotypes for different traits behave separately from one another. Put into fancy science language, the law states that each pair of alleles will segregate independently of other pairs of alleles during meiosis. The pair for pea color will separate independently of the pair for pea skin texture.

Bio Basics

You should note that this law doesn't apply necessarily to allele pairs that are on the *same* chromosome. If allele pair A, for example, sits right next to allele pair B on chromosome 2, obviously they're more likely to assort together wherever that chromosome 2 goes. But the traits Mendel examined all behaved like traits with genes on *different* chromosomes.

Punnett Practice

Go back to the Punnett squares figure. What if Bob were YY instead of Yy? What would his offspring with Doris look like then? What if he were yy instead of Yy? What would their offspring be then? How would the phenotype and genotype ratios look in each instance? All of these crosses involving one trait are called monohybrid crosses.

I'll hint at a few answers here: If Bob were YY, all of his offspring would receive a Y allele. It's all his gametes will carry. So they'll all be yellow because they'll all have at least one copy of Y. If Bob were yy and Doris were still Yy, Bob could only make gametes carrying y. Thus, half of their offspring would be Yy, or yellow, and half would be yy, or green.

Some other traits that Mendel examined were flower color (purple or white, P or p) and pea shape (smooth or wrinkled, S or s). What happens if you predict the outcome for two traits at once? For example, what if Bob is Yy for color and Ss for shape? What are his possible gamete phenotypes? If you said YS, Ys, yS, and ys, you're right.

If Doris has the same YySs genotype, what will her gametes be? The answer is, the same as Bob. Now—what will happen if you cross Bob's possible gametes with those of Doris? For the answer, see the large Punnett square in the figure. This kind of cross involving two traits is called a dihybrid cross.

For a cross involving two traits when both parents are heterozygous for each trait, the offspring phenotype ratio is 9 smooth yellow: 3 smooth green: 3 wrinkled yellow: 1 wrinkled green (9:3:3:1).

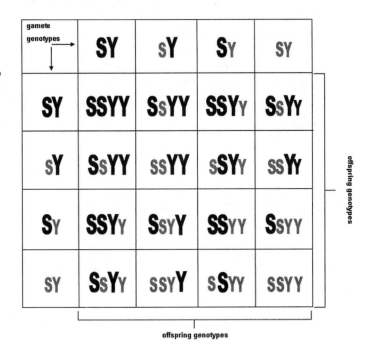

The Probability Question

Refer back to the Yy × Yy Punnett square (Yy Bob and Yy Doris). What is the probability the offspring of these two individuals will be yy? The Punnett square answers that question for you. Only one of the four is yy, so the probability is 25 percent. But some crosses are not quite so easy to calculate. For those, we have a couple of rules.

AND: Multiplication Rules

Looked at another way, the question I just asked is, "What is the probability that an offspring of our Yy Bob and our Yy Doris will receive a y from each parent?" In this situation, we're asking what we call an AND probability question: What is the probability that the baby plant will receive a y from Bob *and* a y from Doris?

When you see an AND question in genetics, you multiply. Each event has a one-half or 50 percent probability, and one half times one half = one fourth. Just as the Punnett square predicted, our little baby plant has a one-fourth or 25 percent probability of getting a y from Bob AND a y from Doris.

OR: Addition Rules

How about an OR question? Let's do this one with dice. You're about to roll a die. What are the odds that you'll get a six or a four? The odds that you'll get a six are 1 in 6 for the standard six-sided die. (This is not a role-playing game die.) The odds are the same for getting a four. Because our question was an OR question, we *add* these two odds together: one sixth plus one sixth = two sixths.

Examples of Mendelian Inheritance

So far, we've just talked about Mendelian inheritance and plants, but lots of organisms exhibit this form of inheritance. Let's be anthropocentric and take a look at some human inheritance. First, some terms.

Dominant vs. Recessive

When the mere presence of an allele in the gene pair results in the phenotype for that allele, we call that allele dominant. When two copies of an allele are required to produce the phenotype, we refer to that allele as being recessive. Thus, in our plants, yellow (Y) is dominant, and green (y) is recessive.

If each allele in an allele pair is the same, we say that the organism is homozygous (*homo* = same) for that trait. Thus, a YY plant is homozygous dominant for yellow peas. If we look at Bob or Doris and their Yy genotypes, however, we'd call them hetero-zygous (*hetero* = different) because they have one of each allele in their gene pair.

Pedigrees and Inheritance Patterns

In humans, some traits exhibit this dominant-recessive relationship, and we can pic-ture those graphically using pedigree charts. These charts give geneticists a visual way to look at traits in families. The following figure shows an example.

In pedigree charts, circles represent females and squares males. Vertical lines indicate offspring. A shaded shape indicates that the individual has the phenotype in question. In this chart, the gene for this phenotype may be recessive, as it stayed hidden, or recessed, until the third generation.

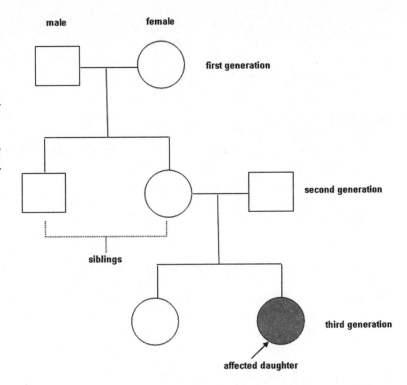

In a pedigree chart, a phenotype present in several individuals of both sexes in multiple generations indicates an association with a dominant allele. Examples of such dominant inheritance in humans include achondroplastic dwarfism and Huntington's disease. Examples of human disorders that are homozygous recessive—requiring two recessive alleles to manifest—are cystic fibrosis and Tay Sachs disease.

Mendelian Muddle: Special Situations

You may have looked around you at your family or another family and thought, but he has red hair, hers is blonde, his is brown … how does that fit into dominant-recessive? Well, not all genes have that clear relationship. Let's take blood type, for example.

There are three possible alleles for blood type, so it's an example of what we call multiple alleles (more than two possible alleles for the trait). The alleles are A, B, or i. The A and B refer to different types of "self" protein flags our cells can stick out of the membrane to let the body know that they are "self." A person with Type A blood uses the Type A flag on its cell; a Type B person uses the Type B flag. Type O people, genotype ii, don't fly any flag at all.

A is considered dominant when paired with i, so an Ai person is Type A. B with i is also considered dominant, so Bi is type B blood. These are dominant-recessive relationships.

But blood type also offers an example of co-dominance, in which the phenotype for each allele in the pair is expressed. A person who has the A allele *and* the B allele has type AB blood. The reason is that people with AB blood fly both A flags and B flags, co-expressing both phenotypes. If an allele pair together produces an intermediate phenotype, say a pink offspring from red and white parents, we say that they show incomplete dominance.

Chromosomes and Inheritance

Inheritance can be bigger than the gene. Sometimes, it has to do with the entire chromosome. In a few instances, a person can develop and grow even with fewer than 46 chromosomes. Conversely, a person with more than 46 chromosomes can, in some cases, also develop and grow. And then there are genes that are specifically linked to the X chromosome, which becomes a special case of inheritance called X-linked inheritance.

X-Linked: Calico Cats and Other Tales

In mammals, males usually have one X chromosome, while females have two. The X is a huge chromosome with tons of genes on it. For most of the genes on the X chromosome, one copy is enough. So female mammals usually shut down one copy of their paired X chromosomes in each cell. This shutdown is allegedly random, and the calico cat illustrates the outcome.

The cat fur color gene is on the X chromosome. Say a female calico gets one X with an allele coding for orange and another X with an allele coding for brown. Wherever her cells shut off the orange X, she'll have brown fur. Wherever they've shut off the brown X, she'll have orange fur.

Males have to deal with the fallout of having only one X chromosome because there is no backup. The X chromosome carries, for example, an important gene for a blood-clotting factor. If the code for this gene is mutated and the resulting protein doesn't work, the outcome can be a blood-clotting disorder called hemophilia. A woman can have one bad allele and be just fine as long as her second X allele is normal. But if a male inherits an X chromosome carrying the mutated allele, he will develop hemophilia because he has no backup healthy allele.

Arm-in-Arm: Other Linked Genes

I mentioned previously that Mendel's law of independent assortment doesn't necessarily always apply. The exception happens with linked genes, genes for different traits that are on the same chromosome. We consider genes to be linked when they occur near enough on the same chromosome to follow each other around in meiosis. The closer the genes are on the chromosome, the more linked their behavior will be.

Trisomies and Monosomies

In relatively rare cases, a person can have one too many or one too few chromosomes—more than 46 or less than 46—and develop and grow. When a person lacks a single chromosome per cell, we call that a monosomy (*mono* = one; *somy* = body). When a person has a third, or extra chromosome instead of a pair, we call that a trisomy (three bodies).

The only autosomal trisomy that generally survives fetal development is trisomy 21, also known as Down Syndrome. Because chromosome 21 is so small with a comparatively small number of genes, having three copies instead of two is not always lethal. A person also can have more than one Y chromosome (XYY) or more than two X chromosomes (XXX) or three sex chromosomes in general (XXY) because the extra X chromosomes will shut down.

One way a trisomy can arise is when chromatids fail to separate in meiosis II. The gamete receives two copies of the chromosome, instead of one. At fertilization, another copy is added in for a total of three. A monosomy is also a failure of separation during meiosis, except in this case, a gamete gets no copies of the chromosome. The only monosomy that allows survival of fetal development is having only one X chromosome, rather than two.

Epistasis: Inheritance Above the Gene

Sometimes, one allele pair interferes with another in a phenomenon called epistasis. In epistasis, two or more gene pairs may interact to determine a phenotype. Thus, a mouse might inherit a pair of alleles that should make it brown (Bb), but elsewhere, on another chromosome, is a pair of alleles (Ww) that blocks the Bb effect and results in white fur. In this case, the mere presence of a single dominant W interferes with the Bb outcome.

Mitochondrial Inheritance: Thanks, Mom!

You may think everyone's a 50:50 split of DNA from their biological mother and father. The reality is that you carry a weensy bit more DNA around from your mom.

Remember how females make one big fat egg and three polar bodies? All the mitochondria we have come to us from that big egg. That means that all of our mitochondrial DNA comes from mom, whether we're male or female.

Bio Bits
A few years ago, researchers uncovered a 9,000-year-old male mummy from a cave in Cheddar, England. They sequenced the mitochondrial DNA from local volunteers and actually found three who shared mitochondrial DNA sequences with the 9,000-year-old relative. Evidently, his people like to stay put.

The Least You Need to Know

- ◆ Mendel's work yielded two laws related to gene/chromosome behavior: the Law of Segregation and the Law of Independent Assortment.

- ◆ Punnett squares predict the gametic and offspring genotypes of a mating pair.

- ◆ Inheritance can be dominant-recessive, co-dominant, X-linked, or complicated by other genes and environmental factors.

- ◆ For Punnett squares, we use multiplication to address "*AND*" probability questions and addition to address "*OR*" probability questions.

- ◆ Only a few forms of chromosome number other than 46 survive fetal development in humans.

- ◆ Because we inherit our mitochondria from our mothers, we carry relatively more "mom" DNA around with us.

DNA

In This Chapter

- ◆ The history of DNA's discovery
- ◆ Replication
- ◆ Mistakes in replication
- ◆ Mistakes and evolution

DNA, as the central molecule of heredity, is key to many aspects of our lives (besides, obviously, encoding our genes). Medicines and therapies are based on it. TV shows and movies practically feature it as a main character. We profile it from before we're born until after we die, using it to figure out what's wrong, what's right, what's what when it comes to who we are, and what makes us different and the same.

But it wasn't so long ago that we weren't even sure that DNA was the molecule of heredity, and it was even more recently that we finally started unlocking the secrets of how its genetic material is copied for passing along to offspring.

The History and Romance of DNA

The modern-day DNA story is dynamic and fascinating. But it can't compare to the tale of the trials, tribulations, and downright open hostilities that accompanied our recognition of its significance.

Griffith and His Mice

Our understanding started with mice. In the 1920s, a British medical officer named Frederick Griffith performed a series of important experiments. His goal was to figure out the active factor in a strain of bacteria that could give mice pneumonia and kill them.

Bio Basics

You may think that a good science experiment is a complex brain teaser that only a highly trained researcher could understand and devise. Yet classic science experiments have one feature in common: simple elegance—clean and straightforward experimental designs that directly address an important question.

His bacteria of choice were *Streptococcus pneumoniae*, available in two strains. One strain infects and kills mice and thus is pathogenic, or disease causing. These bacteria also have a protein capsule enclosing each cell, leading to their designation as the smooth, or S, strain. The other strain is the R, or rough, strain because it lacks a capsule. The R strain also is not deadly.

Wondering whether or not the S strain's killer abilities would survive the death of the bacteria themselves, he first heat-killed the S strain bacteria. (Temperature changes can cause molecules to unravel and become nonfunctional, "killing" them.)

He then injected his mice with the dead germs. The mice stayed perky and alive. Griffith mixed the dead, heat-killed S strain bacteria with the living, R strain bacteria and injected the mice again. Those ill-fated animals died. Griffith found living S strain cells in these rodents that had never been injected with live S strain bacteria.

With dead mice all around him, Griffith had discovered that something in the pathogenic S strain had survived the heat death. The living R strain bacteria had picked up that something, leading to their *transformation* into the deadly, pathogenic S strain in the mice. It was 1928, and the question that emerged from his findings was, "What is the transforming molecule?" What, in other words, is the Molecule of Heredity?

def•i•ni•tion

Transformation, Griffith's word for this transfer, is now part of the biological lexicon. In this context, it means a change in genotype or phenotype resulting from a cell's taking up external DNA. Bacteria do this a lot. We, on the other hand, do not.

Hershey and Chase: Hot Viruses

A fiery debate tore through the ranks of molecular biologists and geneticists in the early twentieth century, arguing about whether proteins or DNA were the molecules of heredity. The protein folk had a point. With 20 possible amino acids, proteins offer far more different possible combinations and resulting molecules than do the four letters (nucleotide building blocks) of the DNA alphabet. Protein advocates argued that the molecule with the most building blocks was likely responsible for life's diversity.

In a way, they were right. Proteins underlie our variation. But they were also fundamentally wrong. Proteins differ because of differences in the molecule that holds the code for building them. And that molecule is DNA.

Our first critical demonstration of this fact came thanks to yet another beautifully elegant experiment involving a blender, some viruses, and some bacteria. The scientists who worked the blender buttons were Alfred Hershey and Martha Chase. Like Mendel and Griffith, they chose a simply perfect biological model system to test their ideas.

Bio Basics

Another scientific buzzword is the concept of models. As with using supermodels to test reception of a new fashion idea, scientists use lab models to test out new science ideas. Sometimes, these models are cells in a petri dish, with scientists exposing them to different substances to test their responses. And sometimes, as with Griffith's experiment, the model is an organism, like a mouse.

A group of viruses, called bacteriophages, infect bacteria and use them as viral incubators to make more viruses. Bacteriophages consist of a bit of DNA packaged into a coat of protein. We now know that the protein coat can latch onto the outside of a bacterial cell (such as an *Escherichia coli* or *E. coli*, infamous for its presence in poop), looking much like a lunar lander alighting on the moon. From there, the phage injects its DNA into the bacterium, where it hijacks the living cell's machinery to make more protein and DNA for more phages.

To address the DNA versus proteins debate, in 1952, Hershey and Chase took advantage of two key differences between the molecules, related to their content. Both contain nitrogen, carbon, and oxygen. But only DNA contains phosphorus (in the phosphate of the nucleotide). And only protein contains sulfur (a component of a specific amino acid called cysteine).

With this difference in mind, Hershey and Chase generated phages that had incorporated either radioactive ("hot") phosphorus or hot sulfur. Viruses with the hot phosphorus had hot DNA, while those with hot sulfur had hot protein coats.

Enter the blender. The researchers blended each of their groups of labeled viruses with *E. coli* bacteria and used radioactive detection techniques to see which blend had hot bacteria.

The virus injects the DNA into the target bacterium. With viruses labeled with radioactive (hot) sulfur, the hot material stayed outside the cell. With hot phosphorus, incorporated into DNA, the hot material entered the cell, identifying DNA as the genetic material.

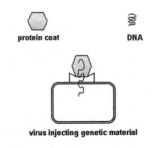

protein coat   DNA

virus injecting genetic material

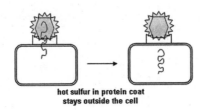

hot sulfur in protein coat
stays outside the cell

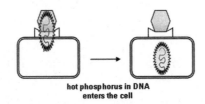

hot phosphorus in DNA
enters the cell

They found that when they mixed the hot-protein viruses with the bacteria, the "hot" stayed outside the cells, even as the bacteria were infected and made more viruses.

But when they mixed the hot-DNA viruses with the bacteria, the "hot" entered the cells. With their results and those from a few other research teams, it was official: DNA was formally designated the molecule of heredity.

Watson and Crick (and Franklin and Wilkins)

From that conclusion, the race was on to figure out exactly how DNA worked as the molecule of heredity. And at this point, things got a bit down and dirty.

Now everyone wanted to figure out how DNA could be duplicated and passed from one organism to another, from parent to child. The way to answer that question was to assemble its components in a way that would allow copying. The work of British research scientist Rosalind Franklin was central to resolving these questions. In the end, however, James Watson and Francis Crick walked into the sunset of fame.

Accounts differ, but the reconciled basics of the scientific soap opera appear to be as follows: Rosalind Franklin was taking pictures of DNA using a technique called x-ray diffraction. She knew what she was doing and what she was seeing, but she also was cautious and wanted to obtain more data before building any DNA models.

> ### Bio Bits
>
> With x-ray diffraction, a researcher targets a crystal with x-rays to produce a three-dimensional picture of it. The rays will scatter, or diffract, at different angles based on the molecule's shape. Film picks up the scatter as dark markings, allowing inference of its shape and structure, like the reverse of reconstructing your own outline using your shadow.

Maurice Wilkins was a colleague of Franklin's, but the two experienced tremendous tensions. Franklin began a move to a different research institution. While details are a bit murky, at that point, Wilkins appears to have shown Watson one of Franklin's x-ray diffraction photos of DNA—known as diffraction photo #51.

Watson, who had seen Franklin present her work two years earlier but had not understood its implications, now saw something more in this picture. In it, he and Francis Crick clearly saw the structure of DNA. And they saw how DNA could copy itself.

Bio Bits

Rosalind Franklin's name was notably absent when the team of Watson, Crick, and Wilkins was awarded the 1962 Nobel Prize in physiology and medicine for their work. The prize can be awarded to only three recipients at once, and it is not awarded posthumously. Franklin had died in 1958 of cancer.

The paper they wrote as a result—one of the most famous papers in science—is a single page published in the top scientific journal *Nature*. Franklin and Wilkins get a bare mention, but they did publish papers in the same issue.

Bio Basics _____

There may be no better illustration of scientific dryness than this quote from Watson and Crick's seminal 1953 paper published in *Nature*, reporting their findings about DNA structure: "It has not escaped our notice that the specific pairing we have postulated immediately suggests a possible copying mechanism for the genetic material."

What Watson and Crick saw and understood about that photograph was based on tiny molecular measurements. The picture and their own understanding told them that DNA was a double-stranded molecule, a zipper, twisted into a spiral helix. The teeth of the zipper were the bases of each strand, reaching across and bonding with one another. As indicated in Franklin's famous picture, the teeth were exactly the length of two bases bonded together down the center.

Watson and Crick hadn't only figured out DNA's structure. They also had determined how DNA copies itself. The zipper would unzip, leaving two strands of DNA with open bases, or teeth. Then the cell could use the old strands to build two semi-new, complete zippers. This process of copying, given briefly as DNA→DNA, is called replication.

Bio Basics _____

The nucleotide bases A and T, and C and G pair together so well that one researcher, Austrian chemist Erwin Chargaff, developed the concept before scientists even understood the copying mechanism of DNA. He analyzed the DNA content of many species and discovered that across the board, the amounts of adenine were roughly equal to the amounts of thymine, and the same relationship applied between cytosine and guanine. His work led to Chargaff's rule, stating that in DNA, A = T and C = G.

Copying DNA: Replication

The teeth on opposite sides of the DNA zipper bond to each other because of complementary base pairing. In DNA, an adenine nucleotide (A), bonds preferentially with a thymine nucleotide (T), while a guanine (G) preferentially bonds with a cytosine (C). So if the teeth down one side of a zipper are AGGCTT, then the teeth down the other side, bonded with them, will be TCCGAA.

If we unzip our DNA zipper, then AGGCTT is now available for other nucleotides to bond. In other words, it serves as a template for complementary base pairing with a fresh new set of nucleotides. Once that new pairing is complete, we have a semi-new double-stranded DNA with AGGCTT on one side and TCCGAA on the other.

If the other half of our old zipper does the same thing, the result is another semi-new double-stranded DNA. The final result? Two semi-new, double-stranded DNA zippers, each consisting of an old strand and a new strand.

So that's replication, using zippers. Let's look at it now using biological molecules.

The Molecular Players

For each step in the unzipping and copying process, there's an enzyme with a job to do. An enzyme called helicase unzips the helix. Enzymes called *polymerases* add new building blocks, or nucleotides, using the old DNA strands as guides. There's a special process required to make both strands copy in the same direction, and it requires two enzymes, primase and ligase. Primase builds small sequences of RNA called primers. Ligase seals up links between nucleotides, ligating them.

def•i•ni•tion

A **polymerase** is an enzyme that builds polymers of DNA or RNA using nucleotides.

Two-Way Confusion

As simple as all of this DNA unzipping may seem, DNA has a specific issue that complicates its replication. The double-stranded molecule, you see, suffers from a terrible case of antiparallelism.

When we depict DNA, read it in our heads, or type it out using its AGCT alphabet, we have to start somewhere. To start reading in a book, you find the first word.

To determine where to start reading on DNA, you find its deoxyribose. But wait, you may be thinking … *every* DNA nucleotide has a deoxyribose. How can we tell the end of a DNA molecule from the beginning? Here's how.

Imagine a string of DNA that's your size, lying on the floor. You walk up to it. If the first carbon you encounter at the tip of the molecule is the carbon 5 of the deoxyribose, you're at the beginning of the DNA molecule, the top. If you find yourself staring at carbon 3, you're at the end.

DNA "runs" or is read starting at the end with a carbon 5 uppermost and ends where there is a carbon 3 downmost. We actually show these as 5' and 3' in molecular biology, which you can read out loud as "five prime" and "three prime." Thus, we start reading a strand of DNA at its five-prime, or 5', end.

We "read" DNA starting at the end with the 5' carbon on deoxyribose hanging off and end where the 3' carbon hangs off.

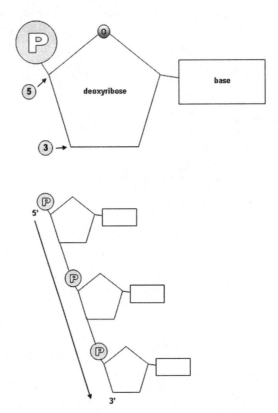

But a double strand of DNA is antiparallel. One strand runs 5'→3' top to bottom, while its partner runs 5'→3' bottom to top! Don't worry—it confuses everyone— including nature.

> **Biohazard!**
>
> Not only do you have to deal with counting carbons in a sugar and knowing your 5' from your 3', but you also have to understand their relationship to upstream and downstream. When your text refers to an element of the DNA occurring upstream, that means it's somewhere 5' to the current location. Downstream means it's in the 3' direction.

The Limitations of DNA Polymerase

The monkey wrench in all this is the intractable enzyme, DNA polymerase. It can add nucleotides by following a template strand in one direction, and in one direction only, adding nucleotides to the 3' end of the new strand.

The problem is, the copying process for both strands at once also takes place in one direction, even though the DNA strands run in two directions. DNA polymerase works great building a new strand that runs 5'→3' in the direction of the copying— the enzyme just scoots right along, building a new DNA strand, nice and smooth. But the polymerases on the other DNA strand, which runs the opposite direction, work away from the direction of replication.

Leading vs. Lagging Strands

The upshot of this complex difficulty is the building of new DNA strands at different speeds. The robotic DNA polymerase quickly builds one new 5'→3' strand in the direction of replication. Because this process happens smoothly and rapidly, we have dubbed this easily built strand the leading strand (1 in the figure on the following page).

That leaves us with the other strand, the lagging strand. The cell plays a little trick on DNA polymerase here to keep the process going the right way, even as polymerase goes the wrong way.

All along the lagging strand, the enzyme primase builds short little sequences of RNA called primers. Unlike DNA polymerase, primase works 3'→5' on the lagging strand, moving in the direction of replication. Every few nucleotides or so, it builds a little RNA lure to attract the DNA polymerase. A polymerase takes the bait (and in doing so follows the direction of replication) and then, once again, diligently and robotically starts building a new DNA strand away from the direction of replication.

This little cat-and-mouse game of polymerase working the "wrong" way even as primase lures it the right way results in a new strand with small sequences of DNA interspersed with RNA primers (2 in the figure). The DNA fragments on this lagging strand are called Okazaki fragments, after the scientists Reiji and Tsuneko Okazaki, who identified them.

Eventually, the RNA primers are removed, DNA polymerase completes the sequence, and the enzyme ligase comes in and closes up the nicks between the fragments. Finally, we have a complete double-stranded DNA molecule.

The limitations of DNA polymerase complicate DNA replication.

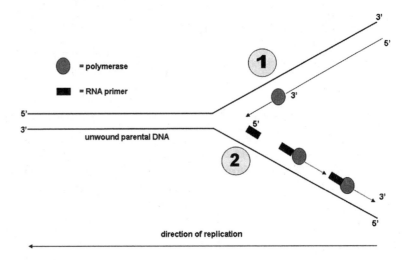

The Big Picture: The Replication Bubble

When helicase unzips the DNA, the resulting structure is like a two-pronged fork, with the point where the prongs meet pointing in the direction of replication. On one prong, the leading strand builds apace, while the lagging strand lags on the other prong.

Let's say that the top prong of our fork has the new strand that runs 5'→3' in the direction of replication (toward the point where the prongs meet). That makes the bottom prong or strand our 3'→5' lagging strand.

And then, there's a bit of a twist. Or at least, there's another fork in the helix. If the fork we've focused on all this time is replicating this way ←, another fork adjacent to it is replicating this way →. Together, the two forks look a bit like a diamond, or <>, which is called the replication bubble. Right down the middle of the diamond is a magical vertical line, the origin of replication, where the two opposite-moving processes start.

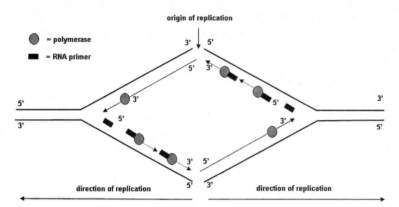

Two forks with replication occurring in opposite directions form a replication bubble. At the center of the bubble is the origin of replication. The leading and lagging strands are on opposite "prongs" of the opposite forks.

When the point of each fork reaches the end of the DNA strand, replication of the entire double-stranded molecule is complete. Because each round of replication preserves one of the two old double strands in each new product, we call replication a semi-conservative process. Every product of replication is semi-old and semi-new.

When Copying Goes Wrong

If you think of DNA polymerase as a typist, you can also realize that this molecule, like any typist, makes mistakes. It's reading the old, template, parental DNA strand and using it to "type" a new strand. What it's really doing, of course, is engaging in complementary base pairing. Thus, if polymerase reads an *A*, it will add a *T* to the new strand. If it reads a *G*, it will add a *C* to the new strand.

But what if it reads an *A* and adds a *G*? Oops. Typo. Can someone get some white-out over here?

Bio Basics

Polymerases, in spite of their diligence, do make errors. In fact, they make an error about every 10^4 to 10^5 nucleotides they put in. Thanks to proofreaders that come along later, the final product will have an error only every 10^9 to 10^{10} bases.

Repair Mechanisms

The cell has proofreaders lined up to follow the polymerase's work and double check it. These proofreaders work in the opposite direction from polymerase, so they "read" the new sequence $3' \rightarrow 5'$. This feature gives them their name, $3' \rightarrow 5'$ exonucleases. Their name also means that they *excise* *nucle*otides, cutting out the incorrect molecule so the right one can be put in. A DNA polymerase inserts the correct nucleotide, and ligase steps in to seal things up.

Mutations, Variation, Evolution

Is a typo always bad? Sometimes, a typo makes noe difference at all in whether or not the code is comprehensible. You likely had no trouble reading the previous sentence, even with the typo in "no." But other times, a typo can result in a completely different cod, with a completely different meaning. If reading that sentence gave you a visual of a fish, instead of DNA code, you can see how a typo that makes a code into a cod makes a difference.

These changes in DNA are called mutations. Just as with our word typos, a mutation can be pretty meaningless and have little effect, and sometimes it can be harmful. Other times, it can be helpful, resulting in a difference for the organism that gives it an advantage in a specific environment. In fact, we'd all still be identical blobs of single-celled protoplasm if it weren't for the mutations that make us all different— from other species and from each other.

The Least You Need to Know

- A few elegant experiments led to the establishment of DNA as the molecule of heredity.

- The race to determine the structure of DNA and how it replicates yielded high drama and a Nobel Prize.

- DNA replication is a semi-conservative process using each old strand of DNA to copy a new strand, with DNA polymerases building the new strands.

- In a replication fork, a leading strand is synthesized on one prong and a lagging strand on the other.

- Because DNA is antiparallel and DNA polymerase works in only one direction, primase builds RNA primers to keep the process moving on the lagging strand.

- Mistakes happen in replication, and although many can be repaired, some persist as mutations, which can be good, bad, or indifferent, depending on their effect.

Chapter 14

Gene Expression

In This Chapter

◆ Using the DNA instruction manual

◆ The three steps of gene expression

◆ Copying the code

◆ Building a protein

The cell copies DNA for one of two purposes: either to make new cells or to make proteins. We covered DNA for new cells in replication in Chapter 13. Now it's time to talk about copying the code so the cell can use it as a guide for building proteins. Until now, RNA has kept a low profile in our discussions. Here it emerges in its starring roles.

DNA→RNA→Protein: The Central Dogma

The discovery of the structure of DNA led to a flood of breakthroughs about how the cell uses the code in the DNA to build proteins, the molecules that do the work of making us all function and giving us our differences. Francis Crick, one of the (many) co-discoverers of DNA structure, coined a telegraphic phrase to illustrate the unidirectional path from

Bio Bits

Francis Crick is probably best known, along with James Watson, for his work in teasing out the structure of DNA. He also made significant contributions to our understanding of how proteins are made, particularly in regard to the mechanisms involving transfer RNAs.

DNA to protein. Because this phrase also encapsulates the focus of most studies in molecular biology—DNA, RNA, and proteins—it is called the Central Dogma of Molecular Biology. Simply put, it is this: DNA→RNA→Protein.

Of course, as with anything in this book, "simple" describes only the surface concept. What goes on underneath is both fascinating and complex. You didn't really expect all the diversity you see around you among organisms to have a simple basis, did you?

Instructions→Copy→Product

In the header for this section, I have rewritten the central dogma without the sciency terms. You may recall in Chapter 5 that I described the cell essentially as a protein factory. As with any factory, there is going to be some paperwork.

In our paperwork analogy, DNA is an instruction manual. The cell isn't building every single protein described in the manual; it's only building some of them, some of the time. So the cell won't rip pages out of its instruction manual and send them to the factory floor. If you were in charge, what would you do? Likely, you'd turn to a copy machine and copy only the instructions you need from the original manual and distribute those instructions to the factory workers. Well, that's exactly what the cell does.

Another consideration is what to use as your copy material. Do you want to use high-grade paper for a temporary document? Probably not. So you copy your instructions using relatively unstable materials that eventually will disintegrate. DNA is quite stable, so it won't do for copying. Your material for copying will instead be RNA, the more unstable but oh-so-useful Other Nucleic Acid.

Once the copy is made, we call it a "message," and we call the molecule the cell uses as the copy "messenger RNA," or mRNA. That message is free to leave the nucleus after a bit of processing. It goes out to the factory floor, where the workers read its instructions and build the protein of interest. Thus, what biologists write as DNA→mRNA→protein, we can also think of as Instructions→Copy of Instructions→Product.

Also Called "Gene Expression"

If you pursue a career in any of the biological sciences or in health or medicine, you may encounter again and again the phrase gene expression. Perhaps you've bumped into it already and wondered to yourself, "What is it that genes express? Anger? Resentment? Contentment?"

Well, while genes can underlie our own emotional expression, their expression is a different construct entirely. When a gene gets switched on—meaning it's been made available to be copied into a temporary message—that's when gene expression begins. Gene expression ends when the final product, a protein, is built based on the code that the gene contained. In other words, gene expression is just another way to say DNA→mRNA→Protein.

Biohazard!

Although the central dogma holds up in general, you should remember that proteins are not the only molecules encoded in the DNA. The code for ribosomal RNA and transfer RNA also resides there. Thus, gene expression, while usually referring to using the code as a map for building proteins, can also refer to transcription of the gene for making these RNAs.

A Two-Step Process ...

If you take the central dogma and break it into digestible bits, it would look like this:

Bit 1: DNA→mRNA

Bit 2: mRNA→Protein

So the first part of gene expression is Bit 1, which we call transcription. The second part of gene expression is Bit 2, which we call translation.

With Many Other Steps

Each of the two big steps of gene expression has three smaller steps. These steps of transcription and translation have the same names: initiation, elongation, termination. As you can see, these aren't fancy terms but words you may recognize and possibly even interpret without my having to write another syllable. Initiation means starting the process. Elongation means making an ever-lengthening molecule. Termination means, well, terminating the process.

While the general events of initiation, elongation, and termination are similar for transcription and translation (starting, lengthening, stopping), the players and the outcomes of each of the two big steps differ completely. So let's take them, one baby step at a time.

Transcription: Copying the Code

Imagine you are a typist whose job it is to place a document on a document stand and then type up a new copy, using the original as your guide. We call the process of typing from an original "transcribing" in the world of office cubicles, and biologists have lifted the term for their own uses. When the cell "types" up a copy of the DNA using RNA for the copy, we call that process transcription.

Bio Basics _____

Transcription is the copying of DNA into RNA, and it's not used only for mRNA. Other kinds of RNA, such as ribosomal RNA, are also encoded in the DNA and copied by transcription.

Transcription works on the same principle as DNA→DNA, or replication: complementary base pairing. There is a twist, however: in replication, or DNA→DNA, an *A* in the template would mean a *T* added in the copy. In transcription, which is DNA→RNA, wherever there's an *A* in the original, the enzymes will add in an RNA *U*, for uracil, to the copy. Thus, a code on the DNA that reads ATG will get copied as UAC using RNA.

Initiation, Elongation, Termination

It all starts with a signal. The signal could come from another cell. It could come from within the cell. The signal says, "We need more of protein X!" (Protein X is a purely fictitious protein created here for the purposes of illustration and is not meant to resemble any real protein, either living or dead.) When the cell gets that message, it transfers its orders to the nucleus. (We discussed how that transfer happens in Chapter 9.)

In the nucleus, the DNA lies wound up tight and unavailable. When the signal arrives that Protein X levels are low and require boosting, things come alive. The DNA relaxes. The transcription factors gather at a meeting place on the DNA called a promoter.

Most genes have their own promoter. It's just another sequence of DNA, but it's an oh-so-important one. It must be available for the molecular machinery of initiation to bind. The initiation machinery recognizes the special promoter sequence on the DNA and binds to it. Once that happens, the gene is considered "on," and transcription has initiated.

When the right proteins are assembled at the promoter, the signal is present for the star worker in the transcription process to enter. This worker is called RNA polymerase. As its name implies, it is an enzyme that builds polymers—long molecules—from RNA building blocks (the nucleotides), using DNA as the template.

The work of the polymerase, as it polymerizes, is the elongation phase of transcription. The celebrity enzyme chugs along, busily grabbing RNA nucleotides and adding them to the growing RNA chain. But it's got to stop sometime, right? How does it "know" when the party is supposed to end?

That's where the termination signal comes in. This signal in eukaryotes is actually a bit of code on the DNA that the polymerase copies into the RNA. This code, the polyadenylation signal sequence, is *AAUAA* in the RNA (which means that in the DNA, it was *TTATT*).

Once the code is in the RNA, the polymerase moves about 10 to 35 more nucleotides further before a set of proteins approaches the RNA strand and cuts the brand-new RNA chain free. This shiny new strand of RNA is now a pre-mRNA, ready for processing. Meanwhile, the RNA polymerase, oblivious of having literally lost its thread, keeps moving on down the DNA, copying more RNA for no particular known reason.

A Garbled Message: Introns and Exons

Gene sequences in eukaryotes are a mess. We keep talking about them as a "code," but each length of code has little nonsense bits in it that don't code for anything. We used to call them "junk" DNA until we realized that they do have real jobs, such as regulating which parts of the sequence are available.

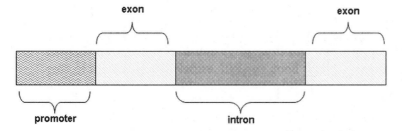

Transcription begins with binding of the transcription machinery to the promoter. Both coding (exons) and noncoding (introns) DNA are transcribed.

The thing is, RNA polymerase doesn't care which parts of the gene it's transcribing are noncoding introns and which parts contain actual code. Introns are interspersed among the coding parts, called exons. Thus, when RNA polymerase is done and the pre-mRNA is released, that sequence it contains needs some serious cleanup. Those introns must go!

Capped, Tailed, Spliced: Messenger RNA

Because the factory workers just do what they're told, they'll build what they read in the code, regardless of whether or not it makes sense. So it's important that the introns get cleanly snipped out of the pre-mRNA before that copy of the code leaves the nucleus.

Luckily, the nucleus is equipped with molecules that know their introns from their exons. This group of molecules, called the spliceosome, clips out the introns. The result is a beautiful mRNA molecule with a smooth, exon-only code.

To prep the pre-mRNA for its exit from the nucleus, the spliceosome clips out the introns.

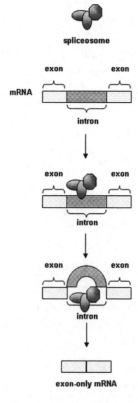

> ### Bio Basics
>
> Introns give the gene sequence flexibility. If the cell chooses, it can leave a bit of an intron in the mRNA, resulting in a completely different protein code. It's like taking a sequence of letters, such as r-a-c-e-c-a-r, as the code, and then choosing parts to make different words. You could take the first four and make *race* or the last three and make *car* or the middle three and make *ace*. With alternative splicing of a coding sequence, the cell can do the same, cutting out a bit or leaving some in to get different uses out of the same coding sequence.

Onward to the factory floor, right? Nope. Not quite yet. Before the mRNA can leave the nucleus through one of the pores, it must put on a hat and tail.

At the 3' end of the molecule—RNA, like DNA, has its 5' and 3' ends—the cell adds on a long string of adenine nucleotides, like this: AAAAAAAAAAAAAAAAAAAAAA, except longer. This tail of nucleotides is called the poly-A tail because, well, it's a lot of *A* nucleotides added on to the RNA's "tail."

Then there's the hat. At the 5' end of the mRNA, the cell adds on a cap of guanine (G) nucleotides. This hat, in another startling exemplar of scientific clarity, is called the "5'-cap."

Now duly spliced, tailed, and capped, the mRNA can proceed on its way out to the factory floor where the workers, the ribosomes, await it in the cytoplasm on the endoplasmic reticulum.

Translation

Have you ever had to read something in a foreign language and figure out what it said in your own language? We call that process, of course, translation. Well, if you're a ribosome, your language is "amino acidese," but the code you receive from the nucleus is in "nucleotidese." Your job as a ribosome is to read the nucleotidese and translate it into amino acidese so that you know which amino acids to use to build the factory's product: proteins.

It may not have crossed your mind to wonder how nature manages to work with only four different kinds of nucleotides to produce the millions of different kinds of proteins cells make. You may also have never stumbled into the question of how to use a code with four letters to designate 20 different words in a different language. But some people—possibly with more time on their hands than regular folk—did ponder that conundrum extensively, wrestling with the seemingly nonsensical math.

They determined how the cell uses a four-letter DNA code to designate 20 different amino acids. Let's follow the reasoning (so easy to do after the fact, right?): If you make the code so that each code word has two letters in it, that's only 16 different code words (4^2 different combinations) for 20 amino acids. No good. But if you kick it up to 4^3 combinations, so that each code word has three letters, then you've got 64 different code words, or *codons*, for only 20 amino acids. Neither one is a perfect fit.

def•i•ni•tion

A **codon** is the three-letter code word for an amino acid.

Perhaps you've heard that some native peoples in cold climates have many different words for snow? The code for amino acids is a bit similar. As it turns out, there *are* 64 codons, each consisting of three letters. But the ratio is not one codon:one amino acid; instead, some amino acids are like snow in cold cultures—there's more than one word to indicate them.

For example, just as in English we can call a popular starchy tuber either "potato" or "spud" and mean potato either way, the amino acid phenylalanine is UUU or UUC in nucleotidese—they both mean "phenylalanine." This redundancy in the code has led scientists to refer to the genetic code as degenerate, a term that conjures up retro attitudes about criminals in the early twentieth century but that here really just means "redundant."

Bio Basics

There is safety in the degeneracy (redundancy) of the genetic code. Because of it, a mutation may not change the amino acid that's encoded. An example is serine, with the code words *UCU, UCC, UCA,* and *UCG.* Any mutation in the third letter will still encode serine. Thus, the redundancy provides some protection against any potential negative effects of mutation.

A few words about four specific and important codons and then we can start building our first protein! The ribosomes receive a messenger RNA that may have a bit of fluff in the beginning, some introductory sequence information that's not really code for the protein to build. How do the ribosomes "know" when to start reading the code? They come across a special codon in the sequence called the start codon. This codon is AUG, and it's the nucleotidese code word for the amino acid methionine (met). The first AUG the ribosomes encounter triggers the real start of translation.

second letter

first letter	U	C	A	G	third letter
U	UUU = phe UUC = phe UUA = leu UUG = leu	UCU = ser UCC = ser UCA = ser UCG = ser	UAU = tyr UAC = tyr UAA: STOP UAG: STOP	UGU = cys UGC - cys UGA: STOP UGG = Trp	U
C	CUU = leu CUC = leu CUA = leu CUG = leu	CCU = pro CCC = pro CCA = pro CCG = pro	CAU = his CAC = his CAA = gln CAG = gln	CGU = arg CGC = arg CGA = arg CGG = arg	C
A	AUU = ile AUC = ile AUA = met **AUG** = met	ACU = thr ACC = thr ACA = thr ACG = thr	AAU = asn AAC = asn AAA = lys AAG = lys	AGU = ser AGC = ser AGA = arg AGG = arg	A
G	GUU = val GUC = val GUA = val GUG = val	GCU = ala GCC = ala GCA = ala GCG = ala	GAU = asp GAC = asp GAA = glu GAG - glu	GGU = gly GGC = gly GGA = gly GGG = gly	G

The genetic code is degenerate, meaning that it is redundant: for some of the 20 amino acids it encodes, the cell uses more than one code word. Because the ribosomes receive the code in mRNA, we use RNA nucleotide symbols to indicate the codons.

AUG also sets the reading frame for translation. Every nucleotide sequence following AUG will be read in groups of three, even if there's an error in the code. One such error might be a deletion or insertion mutation, the removal or addition of a single nucleotide. Such a change will shift the reading frame of the sequence by one nucleotide, which can result in a completely different set of codons and a completely different protein.

Bio Basics

The genetic code is the unifying code of life on Earth. Almost every organism on Earth uses this set of triplet codons for encoding and building proteins. The exceptions are small differences, such as stop codons also encoding amino acids in some organisms.

This kind of frameshift mutation is best understood using regular, everyday words. Let's make a sentence of triplet codons that reads *THE FAT CAT ATE THE RAT.* If we remove a single nucleotide, the letter *H* from *THE*, it changes the entire sense of the sentence, read in groups of three letters, as the ribosomes would do: *TEF ATC ATA TET HER AT.* The outcome is a totally different sequence of code words. Frameshift mutations can result from mistakes in copying DNA into DNA during replication or mistakes copying DNA into RNA during transcription.

And of course, translation has to stop. You might think it would just stop when the ribosomes run out of mRNA to read, but that's not how it works. Instead, the 64-word amino acid code contains three words that usually don't code for an amino acid at all (so really, only 61 words indicate amino acids). Instead, these three code words are stop codons. When a ribosome gets to one of these—UAG, UGA, UAA— the whole translation machinery falls apart, and translation ends.

To the Cytoplasm: Let's Build a Protein

Okay, let's now step our way through the protein-building process. The mRNA pokes its way out of the nucleus and goes to awaiting ribosomes, our factory workers, fluent in both nucleotidese and amino acidese.

Ribosomes are made of ribosomal RNA (rRNA) and protein, and they have three specific sites where important protein-building events occur. Each of these sites interacts with another kind of RNA that carries amino acids to the factory workers, called transfer RNA, or tRNA.

A tRNA has two important parts: the part where it bonds to the amino acid it carries, and the part of its RNA sequence that can recognize and complementary base pair with an mRNA codon. Because this three-letter sequence is the complement to the codon sequence, we call it an anticodon. Thus, a tRNA carrying the amino acid methionine, associated with the codon AUG, will have the anticodon sequence UAC.

The tRNA is really an L-shaped nucleic acid with an anticodon in a critical loop to match the codon in the mRNA the ribosomes are reading. This stylized tRNA shows the anticodon for methionone, UAC, to match the codon for methionine, AUG. The tRNA carries the amino acid that matches its anticodon.

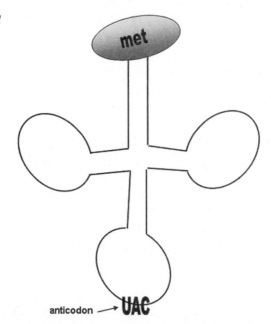

As you can see, we've got three different kinds of RNA we're dealing with here. In fact, if you're keeping track, translation involves an interaction of this RNA trifecta: mRNA, rRNA, and tRNA. Let's see how they work together.

Ribosomes, Meet the Message

The mRNA and rRNA meet up first. The ribosomes start "reading" the nucleotidese message at its 5' end. As described earlier, when the ribosome gets to an AUG start codon, translation really begins, in the initiation step of translation.

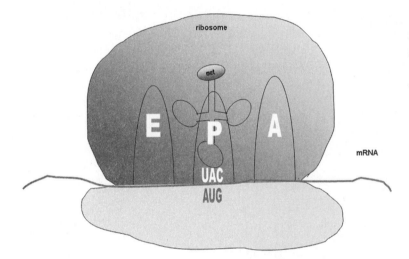

When the ribosomes encounter the AUG start codon, the tRNA with the anticodon UAC occupies the P site, carrying its methionine.

The middle site that the mRNA binds in the ribosome is the *P* site, which stands for peptidyl RNA. It is the site where tRNAs bound to the growing protein chain—or peptide sequence—sit. The first site in the ribosome is called the *A* site, to indicate aminoacyl-tRNA site. This site is occupied by tRNAs still carrying their amino acids to be added to the growing peptide chain.

The final site on the ribosome is the *E*, or exit site. This site is where the tRNA— no longer carrying its amino acid—takes a quick break before heading out to collect another amino acid.

Biohazard!

Don't let protein terminology confuse you. A peptide is two or more amino acids linked together. A polypeptide is a lot of them linked together, usually destined to be folded into the full-fledged version of the molecule, called a protein.

mRNA, tRNA, and rRNA Work Together

Let's build a Very Short Protein. We're going to start with a DNA code, which I've broken up into triplets for easy tracking: 3'→TAC→AAA→ACC→AGT→ACC→5'. When this code is copied into mRNA during transcription, the mRNA will read 5'→AUG→UUU→UGG→UCA→UGG→3'.

When the ribosomes read this code, they first find the start codon, AUG. At this point, a tRNA with an anticodon reading UAC shows up. This tRNA will be toting a methionine.

This first tRNA will occupy the *P* site, waiting for the next tRNA to occupy the *A* site and have its methionine linked to the next amino acid. Let's look at the genetic code table shown previously. Our next codon is UUU. What does that encode? The table shows that it's a code word for phenylalanine. Can you predict which anticodon a tRNA carrying phenylalanine will have?

After initiation with the start codon and methionine, a second tRNA with the appropriate anticodon occupies the A site (top). The two amino acids are linked together (middle). The empty tRNA moves to the E (exit) site, the second tRNA occupies the P site, and a new tRNA with the appropriate anticodon arrives in the A site (bottom).

If you predicted AAA, you're right! A tRNA with the anticodon AAA and a phenyl-alanine will now occupy the *A* site of the ribosome. Its phenylalanine will be joined through an energy-requiring reaction with the waiting methionine, and lo! We have a dipeptide—two amino acids joined together in a Very Short Protein.

Our next codon is UGG, which is nucleotidese for the amino acid tryptophan. The tRNA carrying this amino acid will have the anticodon ACC. As it occupies the *A* site in the ribosome, the tRNA in the *A* site will move to the *P* site, and the tRNA in the *P* site will move to *E*. Our first tRNA, now free of its methionine and the peptide chain, will take a break in *E* site before going off to collect another methionine.

Our last two amino acids, serine (UCA) and another tryptophan, are added. Our final product is a peptide consisting of five amino acids. Congratulations! We have just completed our first expression of a gene sequence.

We've Built a Protein!

At every step of transcription and translation, the cell regulates things, halting prog-ress or encouraging more production. If this regulation happens before the mRNA is made, we call it transcriptional regulation. If it happens after the mRNA is made but before it leaves the nucleus, we call it posttranscriptional regulation. If it happens after translation, we call that posttranslational regulation.

The Least You Need to Know

- ◆ Gene expression involves transcription of DNA into messenger RNA, which ribosomes translate to build proteins.

- ◆ The genetic code consists of 64 triplet codons that are codes for 20 amino acids and 3 stop codons.

- ◆ Transcription begins at the promoter and ends when the RNA polymerase has transcribed the termination sequence.

- ◆ The pre-mRNA undergoes splicing and addition of a poly-A-tail and guanine cap before exiting the nucleus.

- ◆ In translation, mRNA interacts with ribosomes and with tRNAs carrying amino acids and using anticodons for recognition.

- ◆ Translation begins at the start codon, AUG, which encodes the amino acid methionine and sets the reading frame, and ends when the ribosome encounters one of the three stop codons.

Part 4

Evolution and Diversity

Scientists rely on the tenets of evolution without thinking. Drug therapies rely on it, medical treatments use it, agriculture wouldn't be here without it, and neither would you. Evolution lies at the core of our understanding of life, informing how nature shapes species and delineates them. How life itself began remains a matter of informed speculation, but an enormous body of evidence and observation has allowed scientists to build complex and complete family trees for the earth's array of species, past and present. This diversity finds order in our attempts to categorize it, allowing us to grasp its vastness while still maintaining our sanity.

Chapter 15

Darwin, Natural Selection, and Evolution

In This Chapter

- ◆ Context of Darwin's ideas
- ◆ Darwin, Wallace, and natural selection
- ◆ The Modern Synthesis
- ◆ Other mechanisms of evolution
- ◆ The Hardy-Weinberg equilibrium
- ◆ Convergent evolution

Evolution, a change in a population over time, can be a controversial concept, and things were no different when Darwin first proposed his theory of how evolution happens. Since that time, we've identified several other ways by which evolution can occur. Scientists have synthesized natural selection and genetics and worked out a way to identify if evolution is happening in a population.

The Historical Context of Darwin's Ideas

Charles Darwin was born on February 12, 1809, into a society with fixed ideas about the role of divinity—specifically the Christian God—in nature. Darwin's destiny, as it turned out, was to address nature's role in nature, rather than God's. He was not completely comfortable in some respects with that destiny, but this man was born with his ear to the ground, listening to nature's heartbeat. He was born to bring to us a greater understanding of how nature fashions living things.

Yet, he did not emerge into a howling wilderness of antiscientific resistance. Scientists and philosophers who had come before him had posited bits and pieces of what would become Darwin's own theory of how evolution happens. But it required Charles Darwin to synthesize those bits and pieces—some of them his own, gathered on the significant voyage of his lifetime—to bring us a complete idea of how nature shapes new species from existing life.

Catastrophism: Rebuild and Destroy

Leading up to Darwin's major epiphany, people had started to look around them and face the fact of the fossil record. Their problem was feeling compelled to reconcile the existence of fossils with what they divined from Genesis: that the Christian God had made all the animals that exist. End of story.

To get around the obvious fact that the world no longer contained dinosaurs or trilobites, some scientist-philosophers proposed the idea of catastrophism. In a religious context, the argument was that fossils existed because God repeatedly made and destroyed the world. That way, everything around us was still the product of God's wholesale work over a young-Earth time period, but the existence of fossils also had a theological explanation.

> **Bio Bits**
>
> One of the great proponents of the scientific aspects of catastrophism was Georges Cuvier (1769–1832), a French naturalist whose best-known scientific contribution was establishing extinction as a fact in the early nineteenth century. Cuvier's work with fossils led him to argue that the earth had experienced several huge extinctions.

Interestingly enough, like many examples of ostensible disagreement between science and religion, this one had a kernel of consensus: the planet has, in fact, undergone at least five mass extinctions, almost complete wipeouts of most kinds of species, followed by an explosion of new species.

Inherit a Suntan? Lamarckian Evolution

Inheritance of acquired characteristics was another concept that arose. According to this hypothesis, an organism would change in response to change. Thus, a giraffe would produce baby giraffes with longer necks because giraffes needed longer necks to reach the high leaves of a tall tree.

In the same vein, it would theoretically follow that were I to go out and get a suntan, for example, and then reproduce, I would pass my suntan onto my children. The reality is, of course, I might pass on my ability to tan rather than burn to my children, not the tan itself.

The big breakthrough with these ideas was that they acknowledged the role of the environment in shaping species. With the hypothesis of the inheritance of acquired characteristics, things are just a bit out of order.

The actual interaction occurs between the environment and existing, heritable features of an organism. Thus, in the competition for the high leaves, animals with longer necks won. They got food, which allowed them to be healthy enough to reproduce and pass on their longer-neck genes to their offspring. Carried to its conclusion, the outcome would eventually be a population of animals with long necks. Like, say, giraffes.

> **Bio Bits**
>
> The father of inheritance of acquired characteristics was Jean-Baptiste Lamarck (1744–1829), a botanist who was the first to use the term "biology" in the modern sense of the systematic study of life. Speaking of systematic, he also was a well-known taxonomist (classifier of organisms) and the first scientist to try to formulate a serious mechanism of evolution.

Uniformitarianism: Is, Was, Will Be

One idea that gained real ground during Darwin's time and that continues to hold that ground today—appropriately enough—is uniformitarianism. Darwin read the work of Charles Lyell, a geologist who wrote a seminal geological book in 1830 called

The Principles of Geology. In this book, Lyell observed that natural law likely does not change over time. If a river picks up a ton of silt and deposits it miles downstream, rivers have been doing that since the beginning of time. And according to Lyell, "time" wasn't something that started a few thousand years ago but was on a geological scale of millennia.

This uniformity of natural law through geological time gave scientists an important set of tools. If nature is busy today doing what nature did yesterday or millions of yesterdays ago, that's our ticket to understanding the past. We can see what nature's up to right now and use that information as a model for what nature was busy with millions of years ago. These ideas helped Darwin shape his own ideas about nature's role in forming new species.

Thomas Malthus: A Bleak Competition

Another writer who was a huge influence on Darwin was Thomas Malthus, a political economist who wrote a treatise on overpopulation among the poor folk of London. While Malthus was trying to argue that God made more poor folk than London needed so that the poor folk would be encouraged to work and not be lazy, for Darwin, there was a different take-home message.

The naturalist, who read the work of Malthus in 1838 "for amusement," latched on to a particular observation that Malthus made, that "the power of population is indefinitely greater than the power in the earth to produce subsistence for man." Darwin looked around him and realized that this statement didn't apply only to people. It seemed to apply to every living thing.

This abundance of organisms set the stage for competition for resources, just as it did for people. And in that competition, there would be winners and there would be losers. Winners, with their winning traits, would pass those onto their offspring, accumulating those "winning traits" in the population, like the animals with the longer necks in the competition for higher-up leaves.

The *Beagle:* A Voyage and an Epiphany

Darwin was well prepared to read his Malthus. In late 1831, he found himself booked on a voyage aboard the *HMS Beagle*, as a conversation companion for the ship's captain. The ship's mission was a survey of South America, although the voyage extended over to New Zealand and parts of Australia, as well.

Darwin took advantage of every minute. He collected, he observed, he sketched, he took notes. He looked at different finch species from the South American mainland and the Galápagos Islands near the mainland and noted that the finches seemed to have characters that fit them to their environments. Even though they all appeared to be related to the mainland species, the island species also were, in some significant ways, different.

For example, one species with huge, deep beaks fed on large tough seeds, while another species with tiny pointy beaks ate small insects. Anyone with an inquisitive mind like Darwin's would ask: why did these finches, which seem to have started out as members of the same mainland species, become different from the mainland population in these different ways? Those of us with duller minds would've gotten stuck right there. But Darwin didn't. His answer to this question was, natural selection.

Alfred Russel Wallace: The Unknown Darwin

Alfred Russel Wallace developed the theory of evolution by natural selection at the same time as Darwin. His road to enlightenment came via his observations on another island chain, the Malay Archipelago. Like Darwin, Wallace was a naturalist savant, and on this archipelago alone, he managed to collect and describe tens of thousands of beetle specimens. He, too, had read Malthus and under that influence had begun to formulate ideas very similar to Darwin's. The British scientific community of the nineteenth century was a relatively small world, and Wallace and Darwin knew one another. In fact, they knew each other well enough to co-present their ideas about natural selection and evolution in 1858.

Nevertheless, Wallace did not achieve Darwin's profile in the field of evolution and thus today does not have his name inscribed inside a fish-shaped car decal. The primary reason is likely that Darwin literally wrote the book on the theory of evolution by means of natural selection. Wallace, on the other hand, published a best-seller on the Malay Archipelago.

Bio Bits

Alfred Russel Wallace may be unknown to you unless you have a passion for biogeography, the study of the geographical distribution of species. His meticulous work in this field has led to his designation as the Father of Biogeography.

Evolution: A Population Changes over Time

We briefly encountered evolution in Chapter 1, and now it's time to really get to know it as a concept. First of all, as a student of science, you should never speak of evolution as a "theory." There is no theory about whether or not evolution happens. It is a fact.

There are, however, theories about *how* evolution happens. Although there are several proposed and tested mechanisms for how such changes occur, the most prominent and most studied, talked about, and debated, is Darwin's idea that the choices of nature guide these changes.

Evolution in the biological sense does not occur in individuals, and the kind of evolution we're talking about here isn't about life's origins. Evolution must happen at least at the *population* level. In other words, it takes place in a group of existing organisms, members of the same species, and often in a defined geographical area.

Biohazard!

We never speak of individuals evolving in the biological sense. The population, a group of individuals of the same species, is the smallest unit of life that evolves.

To get to the bottom of what happens when a population changes over time, we must examine what's happening to the genotypes of the individuals in that population. The most precise way to talk about evolution in the biological sense is to define it as "a change in the allele frequency of a population over time."

Natural Selection: A Mechanism of Evolution

Darwin, who didn't know anything about alleles or even genes, understood from his work that nature made certain choices, and that often, what nature chose in specific organisms would turn up again in the organisms' offspring. He realized that these phenotypes that nature was choosing must be passed on to offspring. The notion of heritability—of a phenotype being encoded in a genotype that can be transmitted to offspring—is inherent now in the theory of natural selection.

Bio Basics

Nature is selecting organisms to pass on the genes they carry. That's what they get to do if they're chosen.

What is nature selecting them to do? In the theory of natural selection, nature chooses individuals that fit best into the current environment to pass along their "good-fit" genes, either through reproduction or indirectly through supporting the reproducer. Nature chooses organisms to survive and pass along those good-fit genes, so they have greater *fitness*.

def•i•ni•tion

Fitness is an evolutionary concept related to an organism's reproductive success, either directly (as a parent) or indirectly (say, as an aunt or cousin). It is measured technically based on the proportion of an individual's genes that are represented in the next generation.

One final consideration before we move onto a synthesis of these ideas about differences, heredity, and reproduction: What would happen if the population were uniformly the same for a trait? Well, when the environment changed, nature would have no choice to make. Natural selection cannot be a mechanism for evolution without a choice. And the choice has to exist already; it does not typically happen in *response* to a need that the environment dictates. Usually, the ultimate origin for genetic variation— for this choice—is mutation.

Biohazard!

Don't make the mistake of saying that an organism adapts by mutating *in response* to the environment. The mutations (the variation) must already be present for nature to make a choice based on the current environment.

The Modern Synthesis: Ernst Mayr

When Darwin presented his ideas about nature's choices in an environmental context, he did so in a book with a very long title that begins, *On the Origin of Species by Means of Natural Selection*. Darwin knew his audience and laid out his argument clearly and well, with one stumbling block: how did all that heredity stuff actually work?

We now know, thanks to Mendel, our understanding of mitosis and meiosis, and modern genetics, exactly how it all works. Our traits—whether winners or losers in the fitness Olympics—have genes that determine them. These genes exist in us in pairs, which separate during meiosis so that one member or the other of the pair is passed to our offspring. When this gene meets its partner at fertilization, a new gene pair arises. It may produce a similar trait or be a novel combination. But this is how nature keeps things mixed up, how individuals pass their fitness to offspring, how the stage is set for nature to have a choice.

> **Biohazard!**
>
> When we talk about "fitness" and "the fittest," remember that fittest does not mean strong. It relates more to a literal fit, like a square peg in a square hole, or a red dot against a red background. It doesn't matter if the peg or dot is strong, just whether or not it fits its environment.

With a growing understanding in the twentieth century of genetics and its role in evolution by means of natural selection, a great evolutionary biologist named Ernst Mayr (1904–2005) guided a meshing of genetics and evolution into what is called *The Modern Synthesis*. This work encapsulates (dare I say, "synthesizes?") concisely and beautifully the tenets of natural selection in the context of inheritance. As part of his work, Mayr distilled Darwin's ideas into a series of facts and inferences.

Facts and Inferences

Mayr's distillation consists of five facts and three inferences, or conclusions, to draw from those facts.

1. The first fact is that populations have the potential to increase exponentially. A quick look at any graph of human population growth illustrates that we, as a species, appear to be recognizing that potential. For a less successful example, consider the sea turtle. You may have seen the videos of the little turtle hatchlings valiantly flippering their way across the sand to the sea, cheered on by the conservation-minded humans who tended their nests. What the cameras usually don't show is that the vast majority of these turtle offspring will not live to reproduce. The potential for exponential growth is there, based on number of offspring produced, but … it doesn't happen.

2. The second fact is that … not all offspring reproduce, and most populations are stable in size. See "sea turtles," above.

3. The third fact is that resources are limited. And that leads us to our first conclusion, or inference: there is a struggle among organisms for nutrition, water, habitat, mates, parental attention … the various necessities of survival, depending on the species. The large number of offspring, most of which ultimately don't survive to reproduce, must compete, or struggle, for the limited resources.

4. Fact four is that individuals differ from one another. Look around. Even bacteria of the same strain have their differences. Look at a crowd of people. They're all different in hundreds of ways.

5. Fact five is that much about us that is different lies in our genes—it is heritable. Heredity undeniably exists and underlies our variation.

So we have five facts. Now for the three inferences:

1. First, there is that struggle for survival, thanks to so many offspring and limited resources.

2. Second, different traits will be passed on differentially. Put another way: winner traits are more likely to be passed on.

3. And that takes us to our final conclusion: if enough of these "winner" traits are passed to enough individuals in a population, they will accumulate in that population and change its makeup. In other words, the population will change over time. It will adapt to its environment. It will evolve.

Other Mechanisms of Evolution

When Darwin presented his idea of natural selection, he knew he had an audience to win over. He began by pointing out that people select features of organisms all the time and breed them to have those features. Darwin himself was fond of breeding pigeons with a great deal of pigeony variety. He pointed out that unless the pigeons already possessed traits for us to choose, we would not have that choice to make. But we do have choices. We make super-woolly sheep, purple carrots, dachshunds, and heirloom tomatoes simply by selecting from the variation nature provides and breeding those organisms to make more with those traits. We change the population over time.

Darwin called this process of human-directed evolution artificial selection. It made great sense for Darwin because it helped his reader get on board. If people could make these kinds of choices and wreak these kinds of changes, why not nature? In the process, Darwin also described a second mechanism of evolution: human-directed evolution. We're awash in it today, from our accidental development of antibiotic-resistant bacteria to wheat that resists devastating rust.

Genetic Drift: Fixed or Lost

What about traits that have no effect either way, that are just there? One possible example in us might be attached earlobes. Good? Bad? Ugly? Well … they don't appear to have much to do with whether or not we reproduce. They're just there.

When a trait leaves nature so apparently disinterested, the genes underlying it don't experience selection. Instead, they drift in one direction or another, to extinction or 100 percent frequency. When an allele drifts to disappearance, we say that it is lost from the population. When it drifts to 100 percent presence, we say that it has become fixed. This process of evolution by genetic drift reduces variation in a population. Eventually, everyone will have it, or no one will.

Gene Flow: Genes In, Genes Out

Another way for a population to change over time is for it to experience a new infusion of genes, or to lose a lot of them. This process of gene flow into or out of the population occurs because of migration in or out. Either of these events can change the allele frequency in a population, and that means that gene flow is another mechanism by which evolution can happen.

If gene flow happens between two different species, as can occur more with plants, then not only has the population changed significantly, but the new hybrid that results could be a whole new species. How do you think we get those tangelos?

Horizontal Gene Transfer

One interesting mechanism of evolution is horizontal gene transfer. When we think of passing along genes, we usually envision a vertical transfer through generations, from parent to offspring. But what if you could just walk up to a person and hand over some of your genes to them, genes that they incorporate into their own genome?

Of course, we don't really do that—at least, not much, not yet—but microbes do this kind of thing all the time. Viruses that hijack a cell's genome to reproduce can accidentally leave behind a bit of gene and voilà! It's a gene change. Bacteria can reach out to other bacteria and transfer genetic material to them, eventually altering the phenotype of the population.

Evolutionary Events

Sometimes, events happen at a large scale that have huge and rapid effects on the overall makeup of a population. These big changes mark some of the turning points in the evolutionary history of many species.

Bottlenecks: Losing Variation

The word bottleneck pretty much says it all. Something happens over time to reduce the population so much that only a relatively few individuals survive. A bottleneck of this sort reduces the variability of a population. These events can be natural—such as those resulting from natural disasters—or they can be human induced, such as species bottlenecks we've induced through overhunting or habitat reduction.

Founder Effect: Starting Small

Sometimes, the genes flow out of a population. This flow occurs when individuals leave and migrate elsewhere. They take with them their genes, and the populations they found will initially carry only those genes. Whatever they had with them genetically when they founded the population will affect that population. If there's a gene that gives everyone a deadly reaction to barbiturates, that population will have a higher-than-usual frequency of people with that response, thanks to this founder effect.

There are two key points to make about evolution: First, a population carries only the genes it inherits and generally acquires new versions only through mutation or gene flow. Second, that gene for lethal susceptibility to a drug would be meaningless in a natural selection context as long as the environment didn't include exposure to that drug. The take-home message is this: what's OK for one environment may or may not be fit for another environment. There are no guarantees in nature.

Hardy-Weinberg: When Evolution Is Absent

With all of these possible mechanisms for evolution under their belts, scientists needed a way to measure whether or not the frequency of specific alleles was changing over time in a given population, or staying in equilibrium. Not an easy job. They found—"they" being G. H. Hardy and Wilhelm Weinberg—that the best way to measure this was to predict what the outcome would be if there were *no* change in allele frequencies. In other words, to predict that from generation to generation, allele frequencies would simply stay in equilibrium. If measurements over time yielded changing frequencies, then the inference would be that evolution has happened.

Quantifying the Gene Pool

The Hardy-Weinberg equation uses an assumption of a pair of alleles for a trait. In a population, the frequency (percentage) of each combination of alleles—homozygous dominant, heterozygous, and homozygous recessive—should add up to 1 (or 100 percent). The equation for each of these conditions is

$$p^2 + 2pq + q^2 = 1$$

in which p^2 = one homozygous state, $2pq$ = the two possibilities for heterozygous (pq or qp), and q^2 = the other homozygous state. Thus, if we're talking about pea color (Y and y), then p = Y and q = y. The equation would be $Y^2 + 2Yy + y^2 = 1$. (The 2Yy represents the sum of the combination of Yy and yY.)

Let's take a scenario: we have 160 plants that are YY, 80 that are heterozygous (40 Yy and 40 yY), and 10 that are yy. Of our 250 plants, 64 percent (or 0.64) are Y^2, 32 percent (0.16 + 0.16) are heterozygous (Yy + yY), and 4 percent (0.04) are y^2. So 0.64 + 0.32 + 0.04 = 1. Y is at 80 percent (0.64 + 0.16), and y is at 20 percent (0.04 + 0.16).

Our population of 250 plants reproduces, and we look at the frequency of these alleles in the next generation. If that frequency stays the same for each allele, then the population has not evolved. But if frequencies change, that's evolution.

Defining "Not Evolving"

So what does it mean to not evolve? There are some basic scenarios that must exist for a population not to be experiencing a change in allele frequency. If there is a change, then one of the items in the list below must not be true:

- Very large population (genetic drift can be a strong evolutionary mechanism in small populations)

- No migrations (in other words, no gene flow)

- No net mutations (no new variation introduced)

- Random mating (directed mating is one way nature selects organisms)

- No natural selection

In other words, a population that is not evolving is experiencing a complete absence of evolutionary mechanisms. If any one of these does not apply to a given population, then evolution is occurring and allele frequencies from generation to generation won't be in equilibrium.

Convergent Evolution

One of the best examples of the outcomes of environmental pressures is what happens in similar environments a world apart. Before the modern-day groupings of mammals arose, the continent of Australia separated from the rest of the world's land masses, taking the proto-mammals that lived there with it. Over the ensuing millennia, these proto-mammals in Australia evolved into the native species we see today on that continent, all *marsupials* or *monotremes.*

Elsewhere in the world, mammals developed from a common *eutherian* ancestor. In spite of this lengthy separation and different ancestry, however, for many of the examples of placental mammals in the world, Australia has a similar marsupial match. There's the marsupial rodent that is like the rat. The marsupial wolf that is like the placental wolf. There's even a marsupial anteater to match the placental one.

def·i·ni·tion

Among mammals, there's a division among those that lay eggs (**monotremes**), those that do most gestating in a pouch rather than a uterus (**marsupials**), and **eutherians,** which use a uterus for gestation (placental mammals).

How did that happen an ocean apart with no gene flow? The answer is natural selection. The environment that made an organism with anteater characteristics best fit in South America was the same environment that made those characteristics a good fit in Australia. Ditto the rats, ditto the wolf.

When similar environments result in unrelated organisms having similar characteristics, we call that process convergent evolution. It's natural selection in parallel.

The Least You Need to Know

- Malthus and Lyell influenced Darwin and Wallace in developing their theory of evolution by natural selection.

- Evolution is a fact and is defined as a change in the allele frequency of a population over time.

- Ernst Mayr distilled Darwin's ideas into five facts and three inferences.

- Other mechanisms of evolution are genetic drift, gene flow, bottlenecks, the founder effect, and horizontal gene transfer.

◆ The Hardy-Weinberg equilibrium describes the factors that must be present for evolution not to happen in a population.

◆ Convergent evolution is natural selection in parallel.

Origin of Species

In This Chapter

- Defining a species
- Biological species concept
- Morphological species concept
- Making new species
- Maintaining reproductive isolation

When Darwin traveled to the Galapagos Islands, he encountered a lot of different kinds of finches now classified as different species. But how did he or anyone else know that these finches, all of which he correctly assumed descended from a mainland population, *were* separate species? Given the human tendency to classify and group things, biologists have struggled for centuries to determine the best ways to define a species. As you'll see, sometimes what's "best" really does depend on the species themselves.

For centuries, we have classified animals, usually using visual cues. But for many biologists, especially those concerned with how to define a species, looks aren't everything, and often, they can be deceiving.

The Biological Species Concept

One of the primary ways that animal biologists classify species is on the basis of who can reproduce with whom. This basis for classification is called the biological species concept. At first glance, it might seem pretty straightforward: either two organisms can reproduce together, or they can't. But as with so many things in nature, nothing is as simple as it seems.

To begin with, biologists have a hard time even agreeing on how to define the biological species concept. Some of our most famous evolutionary biologists have taken a stab at it, putting the definition through various permutations until we finally arrived at what is considered the definitive version, courtesy of evolutionary biologist Ernst Mayr: "A species is a reproductive community of populations (reproductively isolated from others) that occupies a specific niche in nature."

def•i•ni•tion

In biology, the role a species plays in the ecosystem is its **niche**.

In this definition, notice the presence of two important operative terms: reproduction and *niche*. Reproduction in this case is sexual, requiring organisms of the opposite sex producing viable offspring that in turn can also produce viable offspring. That latter is important because there are many examples of hybrids of two species that live and breathe wonderfully healthy lives, but that also are sterile. If organisms collectively can't reproduce, they're not really members of that biological entity known as a species.

Mayr also included the idea of niche in his definition. An organism's niche is like its career. It's how it fits into and uses its environment, what its role is among the living and nonliving things around it.

In other words, being able to reproduce together isn't enough. To be considered part of a species, the organisms also must occupy the same niche. In this way, we bring together considerations of a species' ecological responsibilities and its reproductive abilities with other organisms to help define its role in the tree of life.

Bio Basics

Mules are a good example of a perfectly healthy, but almost always sterile, hybrid (of a female horse and male donkey; the hybrid of a male horse and female mule is a "hinny"). They cannot make gametes because the chromosome number of the horse and donkey differ. Thus, attempts at meiosis fail because there isn't an even number of homologues to match up.

This kind of specificity may seem like semantics, but it's important. For example, wolves and the domestic dog aren't the same species. Yet we can breed a wolf with a dog and make a viable hybrid. Does that mean wolves and dogs *are* the same species?

No, because they have very different career paths. A dog is man's best friend. A wolf is a large predator that runs wild in packs and isn't too keen on being domesticated. This distinction is important because without the niche caveat, we might be forced to consider wolves and dogs as members of the same species, given that they can hybridize without our help and produce perfectly viable offspring.

On the flip side, we've got poodles and Great Danes. Even though they look pretty different superficially, they can reproduce together and also occupy the same niche. They are, in fact, members of the same species.

Perhaps you've already spotted some of the drawbacks related to using reproduction and niche as a way to delineate one species from another. Some drawbacks are significant enough to render this approach completely inapplicable to many organisms. But it's also got its benefits, especially in the world of animals.

Applications: Sex Required

Among animals, this species concept works pretty well. There are exceptions, but for animals that reproduce sexually, having the criteria of viable offspring and similar niche winnows out most false-positives in the "are you my species?" competition.

You may wonder why it's so important to figure this stuff out at all. One reason is to get a sense of a species' history by determining what separates it from other, similar species. Where did it diverge from its closest relatives? What makes it different enough to be a different species? How did the speciation event happen?

Shortcomings: Fossils Don't Reproduce

And that takes us to the problems with the biological species concept. Possibly one of the obvious ones is that it usually applies only to existing species. If species don't exist together in time, we can't use the biological species concept to determine who belongs where. Thus, this way of classifying species cannot be used, for example, with fossils.

Or with us. Our closest relatives in the *Homo* lineage were either *Homo neanderthalensis* or *Homo erectus*. Were they really different species? Or were they enough like us that we could have interbred to produce viable, reproducing offspring? We can't answer these questions using the biological species concept because time now separates us from these long-lost relatives. We might be able to answer them in other ways, which we talk about later in this chapter.

Space is another consideration. All of the almost seven billion people on the planet are members of the same species. Yet, we're distributed in many populations globally, some of them geographically isolated until very recently. How can we demonstrate in other populations separated by space whether or not they can reproduce together?

In addition to the problem of separation in time and space, the biological species concept has a few other disadvantages. It simply doesn't apply to asexually reproducing species or to organisms that can self-cross. That cuts out bacteria, lots of plants, some invertebrates, and even a few vertebrates from consideration.

Another problem is that the mere ability to reproduce with something else isn't necessarily that meaningful as a trait. This kind of trait is considered a primitive, or ancestral, trait, one that might be shared among several species with a common ancestor.

As a clearer example of a primitive trait, consider fur among mammals. If we consider only mammals as a group, fur is a primitive trait, one that we all share with the common ancestor that gave it to us. That makes us all mammals, but it does *not* make us all members of the same species. Reproductive compatibility can be similarly common and meaningless, as with dogs and wolves. Mayr addressed this issue in part with his addition about niches.

> **Bio Bits**
>
> Sometimes, we can still trace mating events even without the living species. One group has tried to use the timing of divergence of head lice lineages to determine whether their host *Homo sapiens* at some point mated with other hosts like *Homo neanderthalensis* or even *Homo erectus*.

Finally, it's just plain difficult to measure reproductive compatibility. The sheer volume of breeding required to demonstrate unequivocally who could reproduce with whom—or not—is overwhelming. Sure, it's pretty straightforward when you're considering only gorillas and chimpanzees. But what about the hundreds of species of cichlids (a kind of fish) living in lakes in Africa? Such an investigation is practically impossible.

The Morphological Species Concept

So how *do* we tell apart several hundred cichlid species occupying an enormous African lake? With the biological species concept, physical similarity took a back seat to successful mating. In real-world applications, however, physical similarity can work just fine as an often accurate way of establishing species boundaries. If two organisms look a lot alike, the simplest explanation is that they are the same species.

This approach using the similarity of form to establish specieshood is called the morphological species concept. (*Morph* is Greek for "form.")

Applications: Most Things

By using form, you're free of the problems associated with the biological species concept. We can look at fossils representing long-dead animals and use morphology to determine their relationships. We can't mate the fossils, so our best bet is using their forms to decide.

The study of forms also resolves some other issues. Separation in space means nothing if we're comparing forms. And it gets us past the necessity for sexual reproduction. Even with colonies of cloning bacteria, we can use morphology—shape, structure, even metabolic products—to help us identify their relatedness.

Shortcomings: Our Eyes Deceive Us

Wow, you may be thinking. If the morphological species concept is so great, then why did we ever bother with the biological species concept? Well, sometimes appearances can deceive. The evidence of our eyes is not necessarily always accurate.

Because of similar selection pressures, organisms can look the same yet not be members of the same species. Back to our dogs and wolves—these animals can be difficult to tell apart visually.

On the flip side, visual evidence can mislead in the other direction. When an oak tree flowers in the spring, the flower clusters it produces form catkins, furry-looking strings that eventually fall off of the trees. When the caterpillar *Nemoria arizonaria* eats these spring catkins, chemicals in the food trigger development of the caterpillar to look like a catkin. It's the perfect camouflage, rather like taking on the appearance of a pumpkin to hide in a pumpkin patch.

But this species has a second caterpillar emergence in late summer, when there are no catkins. This second brood eats oak leaves and, instead of developing a catkin appearance, develops to look like a twig.

Obviously, if we relied on morphology and examined this caterpillar in the spring and in the late summer, we might assign the two forms as two different species. In this case, morphology would have deceived us. The caterpillar is actually a single species, but is polymorphic: it has more than one form. (Polymorphism occurs when a single species exhibits two or more distinct phenotypes or forms for a given trait or traits.)

There's also the case of mimicry. It is fairly common for one organism to mimic the appearance or behavior of another organism that is noxious or scary. The mimicking organism, while not actually poisonous or dangerous, looks like the one that is. Once again, relying solely on morphology, we might assign both organisms to the same species. In these cases, applying the biological species concept might help resolve the issue.

The final word on species lies in the DNA and proteins. Genetic analysis gives us a measure of relatedness that no reproductive compatibility or similarity of form could provide. Because DNA and protein sequences are, in essence, a form, I've placed them here as part of the morphological species concept. The expectation is that the more alike a sequence is, the more related the organisms are. We expect, for example, humans to have the same number of chromosomes and the same general arrangement of genes on those chromosomes. Our closest living relative, the chimpanzee, has a different number of chromosomes and a small percentage of gene differences. They are obviously a different species from us.

> **Bio Bits**
>
> Could Neanderthals speak? Recent sequencing of the Neanderthal genome has identified the presence of a version of the FOXP2 gene, which is associated with the power of speech in us.

But what about Neanderthals? We have access to some soft tissue from Neanderthal remains, and genetic sequencing will answer some of the questions about their relatedness to *Homo sapiens*.

Geographic Influences on Speciation

So where do species come from? How do we go from a single population to one in which a group of organisms becomes different enough to be a different species? The choices are roughly limited to two possibilities: they can do it because they were isolated from the rest of the population; or they can do it while enmeshed in the original population, but experiencing different pressures that lead them down a different evolutionary path.

Allopatry: Different Country

When something divides a population and the separated groups start to become different, that type of encroaching speciation is called allopatric speciation. (*Allo* is Greek for "different," and *patric* refers to country—it's where our word "patriot" comes from.) These populations are now in different countries. If they experience different selection pressures in these different areas, they may become different enough to eventually become two different species.

How can populations undergo this kind of division? Well, the event can be big, such as a river forming or an earthquake splitting the earth. Or it can be, from our perspective, something small scale, like a rock slide that splits a plant population into two.

Speciation in animals is often allopatric. Darwin's Galapagos finches underwent allopatric speciation when they separated from the original mainland population. On the islands, as they experienced different selection pressures, they became different enough through generations to ultimately be different species.

Sympatry: Same Country

Speciation in organisms occupying the same geographical area, called sympatric speciation, is less common among animals but is something plants can do pretty easily. Plants are not like us. They can, for example, occasionally undergo a doubling of chromosome number without significant bad effects. In plants that can self-cross, an accident of meiosis might increase the chromosome number: a pollen nucleus with double the chromosomes fertilizing an ovule in the same condition would yield offspring with twice the species' normal chromosome number.

> **Biohazard!** _____
>
> Don't get the idea that animals never speciate in sympatry. They do. It's just harder to demonstrate. An example is a pair of goby species living in corals near Papua, New Guinea. They speciated sympatrically as some members of the ancestral population preferentially focused on a single type of host coral for housing.

If the offspring are viable and self-cross, too, suddenly we've got a new species on our hands. This process, called *autopolyploidy* (self-producing an increased chromosome number), is almost strictly the business of plants.

This kind of practically instant speciation arises because the newly polyploid organisms are instantly reproductively isolated from the organisms with the original number of chromosome sets. In this case, the polyploids may look a whole lot like their nearby ancestors, but their inability to reproduce with them renders them a separate species. Here, the biological species concept is obviously the best to apply.

So Lonely: Isolating Mechanisms

Reproductive isolation is a key idea in the conceptualization of species. How does nature go about keeping one species from hybridizing with another? One way to answer that question is to think how you'd do it. Would you wait until they'd met, mated, fertilized, and produced a zygote? Or would you start sooner for better efficiency?

Prezygotic: Ecological, Behavioral, and Mechanical

In most cases, nature doesn't waste resources on producing an inviable hybrid but keeps a zygote from forming through a variety of gambits. Because these inhibitions happen before fertilization and formation of a hybrid zygote, we call them prezygotic isolating mechanisms.

Probably the most obvious way to keep different species from mating is keeping them separated. Thus, even though we can cross a female tiger and a male lion to produce the generally infertile behemoth known as a liger, nature doesn't let this happen. In the natural world, tigers and lions are a continent apart and do not encounter each other. In addition, their niches differ: we may call lions the kings of the jungle, but that designation is more accurately the tiger's. Many tigers live solitary lives in the jungle, while the lion's domain is the savannah.

These ecological and geographical isolating mechanisms are the most obvious way to keep similar species from wasting their time producing inviable or infertile hybrids. But what about situations in which closely related species live near each other, even seeing each other from day to day? In the case of many animals, nature's solution is behavioral isolating mechanisms.

Mating behaviors can be very specific between the male and female of a species. Certain fish require visual and movement cues, male to female and female to male, before mating can take place. If someone makes a wrong move, the appropriate response doesn't take place, and no mating occurs. Birds are known for their intricate mating dances, each move in the ballet an innate response to a specific behavioral trigger. If someone makes a wrong step, then no mating.

If these efforts fail, nature resorts to something more direct: a bad fit. For fertilization to take place, the sperm delivery system must be a fit for the receiving system. If it's not, the result is mechanical isolation.

Bio Basics _____

In snakes and lizards, the male sex organ is a pair of hemipenes (two "half penises"), rather than a single penis. Their conformation can be alarmingly specific, with some looking like four miniature prickly cacti. Obviously, something with that kind of shape must find exactly the right fit to work right. A female with a receiving organ of the wrong shape will be unable to mate with a male bearing quadricacti hemipenes.

This problem of mechanical fit can extend to biochemical recognition between the sperm and egg. If all else fails, there's one last way to keep gametes apart and avoid investing energy in a dead-end hybrid: no sperm-egg recognition. The egg must recognize and accept the sperm for fertilization to occur. Without this chemical recognition, even two organisms of separate species that successfully mate will still not achieve a zygote.

Postzygotic: Unsuccessful Hybrids

The mule and the liger are examples of unsuccessful hybrids. As we've found, these hybrids are often infertile and an energy-investment dead end. No genes make their way into later generations.

Hybrids can fail right after conception, when the embryo takes over and species-specific genetic developmental programs don't mesh. Or the resulting offspring are weak and unhealthy. Or, as we've found, they seem perfectly healthy but cannot reproduce. In any of these cases, the energy investment in reproducing was a complete waste. That's so unlike nature. So why does nature ever allow it to happen?

Because in some cases, hybrids break the rules and prove hardy and fertile and reproductively successful. It's one of the many ways that nature rocks out a new species. Some healthy hybrids include the wolf-dog mix we've already mentioned, which is so robust that some people view it as an example of hybrid vigor, or improved robustness in the hybrid over the parental species. Another common, human-driven example of this kind of successful hybridization is the corn we grow for food crops.

The Least You Need to Know

◆ Organisms can be grouped as a species based on different concepts, depending on the conditions.

◆ The biological species concept applies best to sexually reproducing organisms and fails with fossils.

◆ The morphological species concept works for fossils and asexually reproducing species but can be misleading.

◆ New species can arise in geographically separated populations or in populations occupying the same area.

◆ Nature uses several mechanisms, including ecological and behavioral, to prevent hybridizing.

◆ Hybrids are often weak and/or infertile but occasionally can be more robust than the parental species.

Chapter 17

Origins and History of Life on Earth

In This Chapter

◆ Ideas about how life began

◆ Major events in the history of life

◆ All about fossils

◆ Wipeout: the great extinctions

◆ Human origins and evolution

No one knows how life originated on Earth, but we have some ideas. We're on more solid ground when it comes to some of the big events that happened after life got going. To understand life's history, we rely primarily on the fossil record, which is sketchy in some places when it comes to the species that interests us most, ourselves. Nevertheless, we've got a pretty clear picture of much of the history of our genus, *Homo*.

Geologic Time: Bigger Than the Time You Know

It's hard to wrap our tiny human brains around the concept of geological time. We tend to view a century as a long time, so expanding our comprehension to encompass the 46 million centuries, give or take, that Earth has existed can overwhelm our meager mental circuitry. The best we can do is understand intellectually, if not intuitively, that 46 million centuries is a really really long time and that during that time, incremental processes built what we eventually came to call life on Earth.

A Very Brief History of Geological Time

Name	Periods	Age (millions of years ago)
Cenozoic era	Paleogene, Neogene, Quaternary	65 to present
Mesozoic era	Triassic, Jurassic, Cretaceous	245 to 65
Paleozoic era	Permian, Carboniferous, Devonian, Silurian, Ordovician, Cambrian	570 to 245
Pre-Cambrian, supereon	Encompasses several eras and periods	4,600 to 570

The scale can be overwhelming, but we have divvied it up into bits. Pieces of geological time can start with huge chunks, called supereons, and be incrementally divided from there into, in order of increasing shortness, eons, eras, periods, epochs, and ages. Right now, it's the Cenozoic era, the Quaternary period, and the Holocene epoch. We have no Age for our current timepoint. I guess that mean we are ageless.

Bio Bits

Some folks in the "geological know" have proposed a new epoch, the Anthropocene. They assert that it began when humans started influencing the environment, perhaps with the rise of farming several thousand years ago (hence, the *anthropo*, which means human).

Historic Context for Understanding Life's Origins

There are as many stories of how life originated as there are cultures around the globe. Even in science, the ideas have spanned the strange to the sublime. Whether cultural or scientific, the underlying thread of almost every proposal is a linear process, one that built on previous life, usually starting with or ending with humans as the pinnacle of biological achievement. When we pause to consider how very little time we've been around geologically speaking, this emphasis on *us* is rather ironic.

Until the last couple of hundred years, most explanations for the presence of life have relied on divinity. More recently, scientists have turned to the material world, laying the onus of life's beginnings on everything from deep-sea vents to lightning strikes. Let's take a look at some of what's going on right now.

Current Thinking

Although the triggers remain obscure, most scientists seem to agree that life began incrementally, with some of those Big 4 biomolecules we talked about in Chapter 4 coming together to form the basic structures of life. The steps would have gone something like this: Nucleic acids would have to assemble inside a protective membrane. Next would be a process for the nucleic acid to copy itself to make more nucleic acid.

So basically, what happened first, either separately or simultaneously, was that the elements of life—the SPHONC—came together in large molecules to form balls and strings. Ultimately, we all consist of these strings (nucleic acids) floating inside these balls (plasma membranes). But how did we get from something so deceptively simple to, say, the dung beetle? The first question to address—and one that remains unanswered—is where it all started in the first place.

Life Started on Earth

Well, duh, you might be thinking. We're talking about life on Earth, so it must have started on Earth. And that's what most, ahem, grounded scientists think. We reference the " primordial soup" where life on Earth began, although the ingredients of that soup and even its temperature are matters of hot debate. And in that soup, some energy input—from lightning or UV or heat—brought the soup ingredients together so that the strings and balls could form.

Or—maybe it wasn't lightning at all. Maybe it was little green men.

Life Was Brought to Earth (By Aliens?)

Francis Crick himself—Nobel winner, co-discoverer of DNA's structure, master molecular biologist—concluded that life on Earth must have begun elsewhere and been brought here. By his calculations—and he was not alone in his thinking—there just hasn't been enough time for all the steps required for life as we know it to have happened in the 4.6 billion years since Earth was born or 3.9 billion years since life is thought to have arisen.

Bio Basics

Scientists have thought that an asteroid bombardment 3.9 billion years ago wiped out any nascent life, so most date life's origins to after that bombardment. But recent studies indicate that life could have survived, making it possible that life dates back as far as 4.4 billion years.

Crick didn't suggest that intelligent life necessarily seeded the planet, although some folks ascribe to this idea. Even though we joke about little green men, sober scientists have proposed that some of those macromolecular basics of life may have traveled here on meteors, for example. Still sober but possibly less convincing scientists have gone so far as to propose a concept called directed panspermia, in which, yes, some equivalent of little green men—perhaps the Q?—purposely sent the seeds of life to Earth via interstellar seed pods of some sort. No, I am not making this up.

Bio Bits

Maybe the beginning was both earthly and otherwordly. All life requires phosphorus. Lightning strikes the earth's surface 44 times a second, on average, transferring heat energy that melts sand and soil into glassy slivers called fulgurites. In doing so, lightning rearranges the soil chemistry to produce otherwise rare compounds called phosphites. Meteorites, which also carry this stuff, and bacteria retain the ancient ability to take it up as a nutrient and use the phosphate for organic building blocks.

Big Events in the History of Life

The smallest unit of life is the cell, so life requires something that can form a plasma membrane. And if you think back to our characteristics of life from Chapter 1, you recall that living things reproduce. That requires an ability to copy genetic material into new strands of it to pass along to offspring.

How did nature achieve it all? One of the first steps may have been the formation of fatty bubbles.

Micelles: Fatty Bubbles

Funny thing about soap and detergent. If you drop them into water, the molecules that make them up will hastily orient into those fun little features of bubble baths— the bubbles. These oily bubbles, called micelles, form for reasons similar to why cell membrane phospholipids orient the way they do: the molecules have hydrophobic tails and hydrophilic heads. In water, they spontaneously orient with heads to the outside, interacting with the water, and tails to the inside, avoiding the offensive liquid.

Even simple fatty acids with a head and a single tail do this spontaneously in water. In high enough concentrations, they spontaneously assemble into structures we'd recognize as vesicles.

Of course, the fatty acids themselves have to form. One place this may have happened is in the deep-sea ocean vents, where two necessary ingredients come together: minerals that can provide the energy for rearranging atoms, and gases that provide some of the SPHONC needed to get life going.

If the fatty acids formed in the water as hypothesized and built to high enough concentrations, then their next step would be to turn those heads outward and tails inward. Voilà ... it's Earth's first micelles. Awwww. Cute little things, aren't they?

> **Bio Basics**
>
> In 1953, Stanley L. Miller and Harold C. Urey put methane, ammonia, hydrogen, and water into a closed system (to represent the conditions of early Earth). They then zapped the system with an electric current, meant to have the effect of lightning. The elements rearranged into organic compounds. Some even formed amino acids.

ATP and Ribozymes: Potential Beginnings

But how did we go from an empty fatty bubble to a cell with genetic material? Simple fatty acids flip around a lot.

The flipping may have served an interesting purpose. If a hydrophilic head, happily interacting with the water, happened to encounter something else hydrophilic out there, it might just have interacted with that, too. If that something else was, say, a string of nucleotides, then we've got a gripping scenario right out of Cirque du Soleil: when the fatty acid flipped its head inward, to the center of the vesicle, it might have flipped the interacting string of nucleotides in there right along with it. Sound nutty? Maybe. But it's an attractive hypothesis for how everything got started.

That nucleic acid flying around in the primordial ooze was likely RNA. That's because this versatile little biomolecule not only holds a genetic code (it does that for viruses all the time) but it also can self-replicate. When it does so, it behaves like an enzyme, and we refer to it as a ribozyme. And as you know from Chapter 14, it also serves three roles in the process of translating the code into a protein.

For two reasons, the first nucleotide that nature constructed may have been adenine. First, with a couple of phosphates added (maybe courtesy of lightning!), it's ATP, an almost universal energy currency. Its ubiquity as an energy-currency molecule among living things suggests that it played that role from early on.

Bio Basics _____

RNA is an exception to the rule that all enzymes are proteins.

Second, adenine is the nucleotide we've successfully and most easily synthesized in the lab. The rationale is, if that's the easiest one for us, it must have been the easiest one for Mother Nature. We tend to think that way a lot.

The Rise of O₂

Speaking of us, you may have noticed that we breathe oxygen. But if we had been around when those fatty bubbles started popping up from ocean vents and fatty acids started flinging RNA around, we'd have dropped dead. That's because most scientists agree that levels of free oxygen, if any, were quite low. So where did all of this oxygen come from?

Well, once upon a time, protocells made of micelles and genetic material eventually evolved into another kind of cell, the cyanobacteria. That's not to imply that it was that simple, but eventually, cyanobacteria arose. As we discussed in Chapter 8, these bacteria were and still are significant contributors to photosynthesis. And as a by-product of their activity, oxygen levels began to build.

How do we know that? Because at a certain point in the geological story of Earth's layers, there was iron. And then after a certain point, rust appeared. Rust is oxidized iron, so the implication is that the buildup of oxygen was to blame for the rust. And now, here we all are.

Fossils and the Fossil Record

The kind of fossil you know best is likely one in which rock has replaced the minerals of life. Bone is a good example—over time, the minerals that make up the bone's structure are booted out in favor of rock. The result—a perfect, rocky replica of the bone.

But there are also other kinds. Some are trace fossils, which don't represent some part of the organism itself but instead preserve the traces it left behind. Dinosaur tracks are probably the most familiar trace fossils.

Bio Bits

Among the most famous trace fossils are the Laetoli footprints, made by early ancestors of humans taking a stroll in moist volcanic soil just before the volcano blew its top again and preserved the footprints in ash. Their shape suggests that these ancestors from 3.6 million years ago walked fully upright.

Other kinds of fossils relate to things the animal may have processed in its digestive system or hatched from, such as fossilized poop—known officially as coprolites—or fossilized nests. The most precious fossil of all may be the rare find that harbors soft tissues with DNA or proteins we can sequence.

Biohazard!

You may have heard the pervasive myth that the fossil record lacks transitional fossils. There are, in fact, many, many examples of transitional fossils in diverse lineages. For many groups, the fossil record is beautifully and abundantly complete over long periods of time.

When Trilobites Ruled the Earth

Early on, the only organisms you might have recognized were sponges. In fact, in their day, about 635 million years ago, sponges were the only animal group around.

They dominated shallow seas and grew as big as cars. Imagine a world completely empty of all the animal life you know, except these immobile, enormous sponges, carpeting the ocean floor.

Bio Bits

Some sponges left behind trace fossil evidence in the form of a steroid they produced, which soaked into rocks. Scientists extracted the sponge-specific chemical from ancient rocks, redating the advent of animal life back from 530 million years ago to 635 million years ago.

A little later—well, really about 114 million years later—the first trilobites appeared. Their ranks peaked at 17,000 species, yet they exist no more. They're our example of the comings and goings of huge swaths of species over time.

def•i•ni•tion

An **arthropod** is an animal that has an exoskeleton, usually a segmented body, and jointed appendages. In other words, they are the bugs of the land and sea.

These *arthropods* appeared on the scene about 540 million years ago and stuck around for 290 million years before vanishing. To put this in perspective, the Primate order—our order—may have started about 65 million years ago (dates vary). The *Homo* genus—our genus—has been around for, oh, about 2.5 million years.

Mere babies, we are. While trilobites were scuttling all over the planet with their three-lobed bodies, our order hadn't even risen yet.

Life's Ends: Mass Extinctions

Why did the hugely successful trilobites vanish like that 250 million years ago? Well, they were victims of one of Earth's five mass extinctions.

Of all the species that have ever lived, more than 99 percent are now extinct. Many vanished during one of these five bursts of rapid, widespread extinction. The latest and most famous is probably the one at the end of the reign of the dinosaurs 65 million years ago, the Cretaceous-Tertiary event. The greatest wipeout, however, was the Permian-Triassic event of 251 million years ago, the one that killed off the trilobites, which had actually survived a previous mass extinction. The Permian-Triassic extinction killed off 83 percent of Earth's genera.

What causes these extinctions? Hypotheses vary and include floods, meteors, extreme climate change, volcanic blowouts, and other catastrophes of biblical proportion.

Bio Bits
Are we in the middle of a sixth mass extinction? Some think so.

Human Origins and Evolution

As a human, you've probably been breathlessly waiting for me to dispense with all of these trilobites and bacteria and other unimportant species and get to the most important species of all: *Homo sapiens*.

For all of the tongue-in-cheekiness of that sentence, it is true that nothing seems quite so fascinating or controversial to us *Homo sapiens* as the question of where we came from. In the world of science, the consensus is that we evolved most directly from an ancestor that we have in common with our modern-day closest relatives, the great apes. Most specifically, our closest living relative is the chimpanzee, as attested by unequivocal results of DNA sequencing.

And somewhere back on our family tree, there was a creature that was a little bit monkey and a little bit humany, and from that we were derived. Likely, it was furry and apish and not a whole lot like anyone you know. And it has taken between 5 and 6 million years to get from that to … us.

Hominid Family Tree

I thought about including a figure of the *hominid* (human or humanlike) family tree, but the darned thing changes practically every other week, so Appendix B has links to the latest information. Right now, we appear to have descended either directly or indirectly from a group of proto-human ancestors called the *australopithecines* (*australo* = southern; *pith* = ape). These southern apes dominate the family tree for some time and also dominate as stars of the fossil record; "Lucy," the rock star museum-touring fossil, is an australopithecine.

And somewhere in there, we split from the australopithecines into the *Homo* line. That line also split, branching off into some evolutionary dead ends. The most successful *Homo* species so far has not been us but was *Homo erectus*. This species, our very near cousin that may share ancestry with us through *Homo ergaster*, was around for about 2 million years. By contrast, *Homo sapiens* have been around for 100,000 years, tops.

Further toward the present in the *Homo* lineage, a species dubbed *Homo heidelbergensis* may have split into two lineages, *Homo neanderthalensis* and *Homo sapiens*. There is almost no doubt that we coexisted with and probably fought with *H. neanderthalensis*, likely contributing to its ultimate extinction.

The "Missing" Link

You may have heard people reference the missing link. What is it? Well, back when our lineage split from the great apes, there was that monkey-humany ancestor. The thing is, we don't quite know who that was yet. Whoever it was, right now, it's just the missing link between us and the great apes. But we're closing in with some interesting discoveries dating to between 6 and 5 million years ago of a hominid genus called *Ardipithecus*.

Out of Africa or Multiregional?

No group of scientists may argue more or more loudly than the folks who study human origins. One of their biggest disputes is whether or not *Homo sapiens* evolved within Africa from ancestral species or the ancestral species migrated out of Africa and evolved in separate populations into *Homo sapiens*, maintaining species integrity through migration. Evidence favoring each of these hypotheses—dubbed "Out of Africa" and "Multiregional," respectively—tilts continually, but currently, according to my nonanthropological-expert review, "Multiregional" appears to have the upper hand. I will now duck as anthropologists throw fossilized jawbone bits at me.

The Least You Need to Know

- While life probably arose on Earth, some scientists have proposed that it arose elsewhere and ended up here.

- The big events in the early history of life likely involved formation of macromolecules and micelles.

- RNA may have been the first gene-encoding molecule, and as a ribozyme, it can catalyze its own copying.

- We became oxygenated thanks to the activity of the photosynthetic cyanobacteria.

- The abundant fossil record indicates that the earth has experienced five mass extinctions.

- Our origins trace to a split of an ancestral lineage into apes and us, and our immediate ancestors were a lot like us.

Chapter 18

The Trees of Life

In This Chapter

◆ Taxonomy and organizing life

◆ Showing relatedness with trees and rings

◆ The classification hierarchy

◆ Building family trees

The human mind automatically seeks patterns and tries to fit what it's processing into some recognizable classification. When we can't categorize something, we get pretty uncomfortable. Try it for yourself. The next time you see a person of indeterminate classification—maybe it's hard to tell their age and you're a bartender—see if this little problem bothers you. If it does … well, you're only human. It's this human tendency that leads some of us to devote years of our lives to deciding exactly how to classify the vast diversity of life on Earth.

Taxonomy: The Science of Classification

As you learned in Chapter 16, deciding what to group into a single species category can get us stumbling around over geographical, ecological, and behavioral issues. Once we finally do identify a group as a single species, however, we've designated it as a taxon (plural, taxa), or organisms grouped into a unit on life's family tree. The term taxon can be so specific that it describes only a single species, or it can be broad enough to describe a huge chunk of organisms grouped into a phylum. For example, the species *Quercus bicolor* is a taxon, consisting of swamp white oak trees, but all of the species in the genus *Quercus* also make up a taxon, consisting of all oak trees of any species.

The study of how to describe and group animals together is called taxonomy. In a sense, it's the ultimate human science.

Binomial Nomenclature: Sense in Any Language

Speaking of sense, I was at a restaurant recently and noted that the menu offered up a redfish filet. The mystery was what exactly the restaurant folk meant by "redfish." If you look up the name on those there Internets, you'll find that depending on where you are and who you are, it could be a red version of snapper, drum, salmon, bass, or perch. That's quite an assortment of fish, all referred to casually as "redfish."

Now, had the menu offered up some delicious *Lutjanus campechanus* instead, I'd have known that they were referring to a fish caught off the Gulf Coast that is commonly called red snapper in these parts. That, my friends, is a primary benefit of the system of binomial nomenclature. No matter what regular people call a specific organism, as long as we've got a fancy two-word Latin-based name for it, we'll always be able to clarify what we mean.

> **Bio Bits**
>
> Carl Linnaeus (1707–1778) is the father of taxonomy. We owe to him our successful system of binomial nomenclature. His book, *Systema Naturae*, was hugely influential for the scientists of his time and who came after him.

Take people, for example. In English, we're people or humans. In Spanish, we are *gente*. In German, we are *Leute*. In Swahili—and I had to look this one up—we are *watu*. But in science, we are *Homo sapiens*, no matter where you are or what language you speak. Really clears things up, doesn't it?

So what does it mean? The first word designates the genus, the second level of classification. The second word separates a species from other members of the genus. Thus, if we're looking at *Quercus alba* and *Quercus bicolor*, we're looking at two different species of oak tree, but they're both oaks in the genus *Quercus*. That means that in the great tree of life, they're pretty closely related, too.

Organizing Life

All of these groupings of species have to be fit together somehow in ways that show how they're related. With species in the same genus, that seems pretty straight-forward, as long as you're sure everyone's in the right genus. But then, how do you decide the relationships among the genera in the same family, or how to arrange families within orders? It can get very complicated very quickly. For a quick example, try doing a web image search on the terms "plant" and "taxonomy." The complexity of the family trees that turn up will make you dizzy.

We have different ways to delineate these relationships, but the typical end result is a tree. This tree is at its most inclusive at its base. When there's a branching, that means that the organisms share some commonalities, but are different enough to split. Thus, in the family tree of life, at the root will be some organism that all of the branches share in common, and the closer two branches are to each other, the more closely related they are.

A Tree: Descent with Modification

One of the buzzwords—buzzphrases?—of evolution and speciation is the concept of "descent with modification." It's a Darwinian term, but what does it mean?

Literally, it means that new species emerge from old species through a series of mod-ifications over time. This modification could manifest in many ways. Until recently, we relied on the outward manifestations of these changes to determine relatedness and who sprang from whom. Now, we can also detect modifications at the level of the gene that can tell us the story of descent and family relationships in molecular detail.

Whenever the modification becomes sufficient for descendents to be considered dif-ferent species, the family tree branches. We can look closely at the tree and see all the twigs that represent the millions of species on Earth. Or, we can zoom out and view the tree on a grander scale, with the branches representing the major modifications that separate, for example, insects from people. Either way, we're going to see a tree because descent implies a linear mode of transmission, and modification implies that the line is going to branch here and there.

A Ring: Horizontal Transfer

And that branching kind of model works just fine if you're an organism that relies on the basic old "transmit your genes to offspring" method of doing things. But what if you're a bit … different? What if you're a bacterium that can toss a gene sequence over to another, existing bacterium and change the other guy right there on the spot?

There goes the whole tree concept. When bacteria do this kind of thing, it's not a vertical transfer of genetic material from generation to generation. It's a horizontal transfer of genetic material from one existing organism to another, often with the result that the receiving organism is transformed. In fact, the process is called transformation, and we learned about it in Chapter 13.

To put this into perspective, imagine you walked up to a blonde friend, handed over a copy of the genes that make you brunette, and watched your blonde friend suddenly go brown. Weird, huh? It's modification without descent.

When organisms undergo modification in this way, they don't branch vertically, passing genes from generation to generation over time. Instead, they're linked horizontally, passing genes and undergoing modification in the same generation. The result of this horizontal handoff isn't a tree. If the organisms share a common ancestor, then they're branches from a trunk. If they reach one another's branches across to each other and exchange genes horizontally, they're not branches any more. They've made a ring, as the figure shows.

While vertical transfer from generation to generation yields a branching tree, horizontal transfer among contemporaries results in a ring.

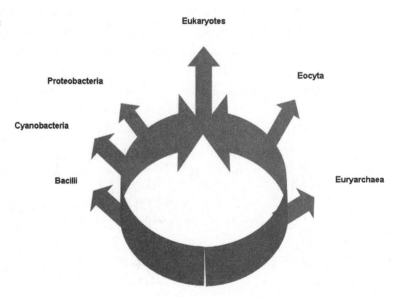

We're going to stick with the tree concept in this chapter, but I do want to point out one significant (to us) ring-forming event in the tree of life. It may be that long ago, some prokaryotes engaged in a bit of this kind of transfer, fusing their genomes horizontally to make something completely unique in the tree—or ring—of life. That something? Eukaryotes.

A Hierarchy: King Philip Came Over For *What?*

We met the classification hierarchy in Chapter 1. The initial letters of each word in the above mnemonic stand for, respectively, Kingdom, Phylum, Class, Order, and Family. The remaining classification categories are Genus (**G**inger) and Species (**S**naps). As you likely understand by now, Kingdom is most general and most inclusive, while Species is most specific and least inclusive.

A note on plants: a reminder that instead of Phyla, the plant folk sometimes use Divisions, so you may have to substitute in King David if you're talking about plants.

Phylogeny: An Evolutionary History

Taxonomy is part of a larger area of scientific focus called systematics. As this name implies, this field involves a systematic look not only at classification of organisms but also at how they are related to each other.

The trees show these relationships, so they are the—ahem—fruit of systematics. Because these family trees are visual representations of descent with modification, they illustrate the evolutionary history, or phylogeny, of the organisms in the tree. Building these trees to accurately reflect a taxon's evolutionary background can keep systematists up at night, worrying about things like parsimony (producing the simplest tree that reflects the facts) and other things that the rest of us have never even heard of. While us nonsystematists sleep soundly in our ignorance, systematists wrestle with the demons of form and function, analogous versus homologous traits, and the many possible permutations of life's family tree.

> **Bio Basics** _____
>
> Trees that we create to illustrate evolutionary relatedness among species are our hypotheses about those relationships. Like any good hypotheses, they are testable and can change with new information.

Using Form or Function

You may be thinking, What's all the insomnia about? Can't you just look at a group of organisms and tell that they're related? Well, sure you can. Sometimes. But sometimes, the eyes can deceive. Take the red panda and the giant panda. They both look kind of panda-ish. In fact, for some time, systematists had them both as panda types, grouped together in a way that indicated a close relationship. Their forms deceived … panda-looking faces, a tendency to eat bamboo in Asia, a certain fuzzy cuteness. And the fun thing about these two animals is that for a long time, in spite of the designation of "panda bear," systematists thought they were both likely related to raccoons.

But the giant panda (*Ailuropoda melanoleuca*) is actually a bear, a member of the *Ursidae* family. In fact, its closest living relative is the spectacled bear, which lives oceans away in South America. The red panda (*Ailurus fulgens*, which beautifully means "shining cat"; you can't say that some taxonomists lack the soul of the poet) is not that closely related to the giant panda and is now considered the sole member of its own family. But their forms and their functions—furry bamboo eaters in Asia—confusticated and bebothered scientists for years.

So how did systematists finally resolve the puzzle? That magical, wonderful molecule called DNA provided the incontrovertible evidence.

Using Molecular Comparisons

Back in the dark ages, when I was an undergraduate taking a physical anthropology class, the anthropologists were up in arms about which of the existing great apes was our closest living relative. Was it the gorilla? The chimpanzee? The orangutan? Battles royale raged through journals and meetings.

> **Bio Bits**
>
> We are so similar to the chimpanzee, from genes on up, that Jared Diamond, the author, geographer, and physiologist, has famously referred to us as the third chimpanzee, the first two being the two existing chimp species, *Pan troglodytes* and *Pan paniscus*.

Then along came DNA sequencing, and now all of the infighting looks rather amusing in hindsight, almost quaint. When we compare DNA sequences between us and the various great apes, the answer is clear: our closest living relative is *Pan troglodytes*, the chimpanzee. We share most of our DNA—98.4 percent—in common. Runner up is the gorilla at 97.5 percent, and coming in a far third is the orangutan, at 96.5 percent.

You'd think that the matter is settled. After all, our eyes can lie, but molecules don't, right? Well, there are two general problems with that assumption. The first is that going by form, rather than molecules, the orangutan has a lot in common with us anatomically and behaviorally. These commonalities of form led some anthropologists back in the Dark Ages of the twentieth century to place the orangutan as our closest relative.

The second issue centers on a problem of analogy. We may conclude that similarities arise from relatedness when they really arise from other factors. You may be taller than average because you had good nutrition as a child. But that doesn't mean that you're closely related to all other tall people in the world. They may be tall because of good nutrition, too. You share the trait not because of a shared gene from a family relationship, but because of environmental experiences in common.

Being of an argumentative bent, some experts still tussle about the relatedness of chimps and humans. Most of the jury, however, has cast its vote in favor and gone home.

Pitfalls: Homology vs. Analogy

Sometimes, organisms seem to have features that make them seem related, but the features are misleading. An obvious example is wings. (Some) insects have wings, birds have wings, and bats have wings. They all have wings because occupying their niche—the sky—meant selection for wings. However, as you can likely discern, a furry mammal, a crunchy insect, and a feathered bird are not actually close relatives in the grand scheme of the kingdom Animalia.

These shared characteristics are really analogous traits: they exist because of similar environmental pressures, rather than because of relatedness. If we built a tree indicating relatedness based on these analogous traits, we'd have one seriously fragile hypothesis. And that's the problem with analogous traits. More subtle analogies can really lead to these kinds of mistakes.

Then there are homologous traits. Homologies arise because of relatedness between organisms. An example is the amniotic egg (the embryo and yolk are surrounded by a tough membranous sac). Animals with amniotic eggs live on land or have an evolutionary history as land dwellers because they had to pack up their egg for life out of water. And the ones that do are clumped together as amniotes. Among the animal kingdom, they're considered fairly close relatives. Their homologous trait is the amniotic egg.

Biohazard!

Do not forget that context can determine whether or not a trait should be considered a symplesiomorphy or a synapomorphy. Fur among mammals is a symplesiomorphy, but among all vertebrates, it's a synapomorphy.

If we compare amniotes with nonamniotes, like fish or amphibians, this amniotic egg is what we call a synapomorphy. Just pause and take that word in. *Syn* = shared. *Apo* = separate. *Morph* = form. All animals that produce amniotic eggs share this form with each other, and the form is separate from that used by other, nonamniote animals. Synapomorphies, also called "shared, derived traits," as long as we identify them accurately as such, are very useful in grouping together closely related organisms.

On the flip side, we have symplesiomorphy. Symplesiomorphies are shared, primitive traits. This is where things get a bit tricky. If you look at the kingdom Animalia as a whole, the amniotic egg is a shared, derived trait, a synapomorphy. We can use it to group together all the animals with it apart from those without it.

Bio Bits

The major take-home message buried in all of these multisyllabic terms is this: in building a tree, synapomorphies help distinguish relatedness, and symplesiomorphies are not that informative.

But … if you look only at amniotes, that amniotic egg is no longer useful for determining relationships because they all have it. Thus, the synapomorphy among the whole animal kingdom becomes a symplesiomorphy among amniotes only. They all have this trait because they all inherited it from a common ancestor. It does not distinguish relatedness among amniotes.

Cladistics: Building a Tree of History

The science of building family trees based on shared evolutionary history is called cladistics. The trees themselves are called cladograms. While we can build trees based on how animals have similar forms or behaviors, cladistics relies on data indicating real evolutionary relatedness.

A Hypothesis of Relatedness

Any cladogram or phylogenetic tree is really a systematist's hypothesis for the relatedness among the organisms it depicts. The human-ape trees I had to learn about in my ancient undergraduate days were hypotheses about the relatedness among humans and great apes. As we have learned, a couple of those hypotheses have now been rejected outright.

Parsimony Principle: The Razor's Edge

A guy back in the 1300s named Occam (variously spelled Ockham or Ockam) allegedly sired a principle that people harken to even several hundred years later: the simplest explanation is always the most preferable. No one in science seems to adhere to that principle, popularly known as Occam's Razor, quite so closely as systematists.

The big rule in building a phylogenetic tree is to strive for parsimony. In the real world, parsimony means being stingy. In the world of building evolutionary histories, it means having as few branches and weird, stray groupings of organisms as possible. Systematists use special computer programs to run their hypothetical trees and identify those requiring the fewest changes and branchings, usually assuming these to be the best hypotheses.

Practice Building a Tree

Building a tree, as you may have discerned, is no easy task. It requires insight, judgment, and considerable knowledge. One of the most important considerations for accuracy about shared and primitive traits is identifying the outgroup.

Take jaws. Not all animals with skeletons have jaws. Let's say we've got a jawless fish called a lamprey and also a salamander, a human, a shark, a mouse, a lizard, and a tuna. What do they have in common?

For one thing, they all have some kind of an internal skeleton, either of cartilage or of bone. So we can't use that to group any of them together. They also all have backbones, so that's another useless symplesiomorphy.

But wait! One of them doesn't have jaws, while the rest do. Yes, we can use jaws as a synapomorphy, a shared derived trait, to group sharks, tunas, salamanders, lizards, mice, and people away from lampreys as being more closely related. Somewhere after the lamprey arose, an ancestor common to all of the other organisms in our list developed jaws.

Because we have other features in common with the lamprey, like a skeleton, but it lacks jaws, we can use it as our outgroup, our example of the ancestral condition we all shared before jaws arose.

> **Bio Basics**
>
> In the mammalian family tree, the outgroup is monotremes, the egg-laying mammals. There are only three classes of monotremes today: two echidna genera and the outwardly risible yet deadly serious platypus.

Can you arrange these animals on the tree based on the presence or absence of specific traits? Note that the lamprey has none of these derived characteristics and thus is the outgroup. Answers below table.

Trait	(a) Lamprey	(b) Salamander	(c) Human	(d) Shark	(e) Mouse	(f) Lizard	(g) Tuna
Jaw?	–	+	+	+	+	+	+
Fur?	–	–	+	–	+	–	–
Amniotic egg?	–	–	+	–	+	+	–
Lungs?	–	+	+	–	+	+	–
Bony skeleton?	–	+	+	–	+	+	+
Bipedal?	–	–	+	–	–	–	–

Answers, from left: a, d, g, b, f, e, c

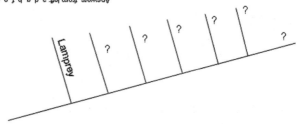

How else could you divide this group of animals? The table in the figure shows who has what. How would you arrange them on the tree? Look at all of the traits and their presence or absence for each species. Who has the most derived traits? The fewest?

The Least You Need to Know

- Taxonomy is the science of classification, and we keep organisms straight using a system of binomial nomenclature.

- Phylogenetic trees are hypotheses of evolutionary relatedness.

- Synapomorphies are informative about evolutionary relationships, but symplesiomorphies not so much.

- The goal in tree building is observing the parsimony principle, producing the simplest tree that reflects the facts.

- Analogous traits, arising in unrelated species because of similar selection pressures, can be misleading.

- In phylogenetics, the outgroup reflects the ancestral condition of the group of organisms under consideration.

Chapter 19

A Survey of Life: Diversity

In This Chapter

- ◆ An overview of domains and kingdoms
- ◆ Protista, the kingdom of confusion
- ◆ A closer look at fungi, plants, and animals
- ◆ Key features of each group
- ◆ Plant and animal diversity

In spite of some of the delicious little mysteries that linger in systematics, many classifications are pretty straightforward and much has been achieved. Now you get to learn about the fruits of this labor in a little stroll through the domains and kingdoms of life.

Domains: An Overview

We've been wrong before about classification. From two kingdoms to six, back down to four plus three domains ... how can we be sure that we've got it right this time? One reason is RNA sequences. Ribosomal RNA doesn't change much over the ages. It stays like it is because it does exactly the same job in any kind of cell, playing its roles in translation.

Any changes that accumulate do so very slowly. For this reason, we can use ribosomal RNA to compare relationships at the deepest points in the family tree, back billions of years.

These comparisons tell us that splitting the earth's organisms among three domains makes a lot of sense. Bacteria likely arose first, followed by Archaea and Eukarya. Throughout their long histories together, representatives of these domains may have swapped genes horizontally, a Bacteria cell handing off sequences to an Archaea, or both kinds of prokaryotes fusing, possibly, to form what would become Eukarya.

But besides the story that lies in rRNA, how can we tell our Bacteria from our Archaea? Well, really, it's more gene sequence comparisons. The more scientists looked, the more they realized that Archaea share as much of their sequence with eukaryotes as they do with prokaryotes. In addition, as we've discovered, Archaea have a predilection for Earth's stranger places. In fact, some scientists think that these archaic organisms may be living representatives of some of the earliest life that bloomed in hot ocean vents.

Eukarya doesn't require gene sequencing to distinguish from the other two domains; a microscope or even the naked eye will do. If it's multicellular, it's Eukarya. If it has a nucleus, it's Eukarya. Yet, although the organisms we likely know best are all eukaryotes, it's the prokaryotes—Archaea and Bacteria—that rule the world. That said, Eukarya is the domain that gets to have the kingdoms: Protista, Fungi, Plantae, and Animalia.

Kingdoms: An Overview

As its name suggests, a kingdom encompasses a variety of organisms grouped together because they have some broad characteristics in common. With something like fungi, that grouping can be pretty straightforward. Some may or may not be quite established, like the proposed Chromista, a developing kingdom of sometimes-photosynthesizing, eukaryotic nonplants that may include some representatives of what we now call algae and fungi. The Chromista kingdom is one of many splits from the one kingdom that's had taxonomists in a tizzy for decades, the Protista.

Protista: A Mess of a Kingdom

This kingdom is a mess because one characteristic—being a single-celled eukaryote—is really pretty much the sole defining trait. If it can't be dropped into fungi, plants, or animals, it ends up in the kitchen sink that is Protista. Protista is such a taxonomic quagmire that some researchers have sought to break it up into as many as 20 kingdoms, without success, to the relief of undergraduate bio majors everywhere.

Why the mess? Because protists come literally in all shapes, and they sometimes have more dissimilarities than similarities. In fact, the best way to determine whether something is a protist is just to determine that it is a eukaryote that is not a fungus, plant, or animal. Some organisms we've dumped into Protista are more closely related to fungi, plants, or animals than they are to each other. Obviously, taxonomists still have their work cut out for them with this bunch.

Because of this confusion, Protista is (a) rapidly losing its status as a real kingdom and (b) really more of a loose grouping of three kinds of organisms that are not officially plants, animals, or fungi. Confusingly, they are generally grouped based on whether they are plantlike (photosynthesizing), funguslike (externally digesting things), or animal-like (everything else). And they represent the vast majority of eukaryotic organisms.

> **Bio Bits**
>
> The cellular slime mold, which is neither a slime nor a mold but a protist, has an interesting and category-defying lifestyle. It can exist as single cells, as colonial cells, or in times of direst need, as a multicellular organism with two tissue types.

Protists might be completely asexual, sexual, or do one or the other. Protists like amoebas might get around by blobbing, while others use cilia, motorized hairlike extensions that allow them to scuttle around. Still others get about using a flagellum. Their shapes are almost infinite, in part thanks to a group called the diatoms, which are single-celled algae with silica armor that take on a huge variety of forms. These have been reclassified by some into the developing kingdom, Chromista.

A few break the single-cell mold and exist in multicellular forms or shift back and forth between the two. Perhaps the most recognizable multicellular protists are the algae, which we sometimes refer to colloquially as seaweed. Green algae does, in fact, live at the edge of our possibly arbitrary division between plantlike protists and actual plants.

Fungi: Decomposing Death

Look around you. Are you up to your ears in dead stuff? Unless you're hanging out somewhere rather unsavory, probably not. And you've got fungus (plural, fungi) to thank for that. This kingdom has some of the greatest names in biology, including rusts, smuts, and puffballs.

Some Major Fungal Groups

Name	Features
Ascomycota	"Sac fungi." Most representatives among fungi, including Baker's yeast and penicillin. Common pathogens, including the agent in thrush. Main feature is the ascus, or sac, where meiosis occurs.
Basidiomycota	About a third of fungus species. Includes mushrooms. Can break down wood. Most have basidia, where spores are produced.
Chytridiomycota	Called chytrids. Found worldwide. Most primitive fungi. Saprotrophs, break down structural proteins and sugars. A factor in worldwide amphibian declines.
Zygomycota	That mold on sugary food. Sexual and asexual reproduction with unique zygospore. Harmless or seriously pathogenic. Agriculturally significant.

Many fungi are saprobes, meaning that they eat detritus, or "dead stuff," although some fungi form friendly relationships with other species and obtain nutrients that way. While we've identified more than 70,000 species, there may be as many as 1.5 million out there fungusing up the planet. What gives an organism membership in the kingdom is external digestion followed by nutrient absorption. They secrete enzymes into the environment and then soak up the digested bits for nutrition. Also, their cell walls are made of chitin, just like an insect's exoskeleton. Mmmm. Crunchy.

> **Bio Bits**
>
> The largest known single fungal fruiting body grew to maturity in the Royal Botanic Gardens in Surrey, UK. It peaked at 625 pounds. Sadly, it met an untimely end when a fox burrowed underneath it, which seems like a very British sort of demise.

While we need them around to take out the trash, we also suffer economically and physically from their effects. Fungi cause devastating plant diseases known as rusts, wilts, and blights, and they also can cause deadly diseases in humans, especially people with compromised immune systems.

Plantae: Immobile Sugar Producers

You've met plants. Their defining features are that they are usually immobile, multicellular organisms that photosynthesize using chlorophyll *a* and *b*, and store up the results as starch. We've counted up about 350,000 species, including green plants, mosses, ferns, liverworts (the world's greatest plant-related word), and seed plants.

Animalia: Anemones to Zebras

If you're a multicellular, heterotrophic, eukaryotic organism engaging in internal digestion (which I'd bet you are), you're probably an animal. That means you, the platypus, the sponge, and the water bear all fit into this kingdom.

Close-Up on Fungi

Let's take a closer look at the fungus among us.

Variety: Ascomycetes to Zygomycetes

Fungi don't eat only dead things. Their roles can also include being parasites or engaging in mutually beneficial relationships with other organisms, as they do with algae to form lichen. They can live as single cells, as yeasts, or in multicellular form, like the shiitake mushroom. Either way, they can really make food tasty. Of course, they can also be deadly.

Life Cycles: Your Gametes Are Showing?

We use the term "yeast" in general when we think of the little critters we use to make bread puff up, but the term really implies something broader. A single-celled fungus is called a yeast. If it's more than one cell, it's a filament. Some fungi can be both, many can be just filaments, and a few are only yeasts.

Lots of fungi have a choice—they can reproduce using sexual or asexual reproduction. For the latter, the process is mitosis. When a yeast

> **Bio Bits**
>
> The largest living organism is a honey mushroom fungus that occupies the space required for 1,665 football fields (2,200 acres). It lives three feet underground in an Oregon forest, could be as old as 7,000 years, and measures 3.5 miles across.

does this, for example, we call it "budding." But a couple of yeasts can also experience an attraction for one another and engage in the yeast version of mating, recombining genes to produce a diploid yeast offspring. Confusingly, the *mating types* of yeast have been designated as *a* and the Greek letter *alpha*.

The mold you see on bread is the haploid form of the fungus. Unlike us, the haploid versions of fungi are the big, visible parts of the life cycle, while the diploid forms remain small and hidden (in us, we're diploid and we keep our haploid gametes tucked away, out of sight). That fungus that is your bread mold consists of cells linked together to form hyphae. These hyphae, if the mood suits, may put out chemical signals that attract another nearby hyphae. That's basically bread mold sex in action. Their nuclei eventually fuse (it can take years for some fungi) and form diploid cells that undergo meiosis, producing spores that can grow again as happy, haploid, bread-digesting hyphae.

When the hyphae first fuse, the outcome is called a mycelium, and the fusion process is called plasmogamy. Fusion results in a truly diploid cell that then undergoes meiosis.

def•i•ni•tion

Responding to chemical signals, haploid yeast cells can alter into either the *a* or the *alpha* **mating type**. Because of the pearlike shape the cells take on, scientists whimsically decided to call them "shmoos," after the fictional animals in an old comic strip, *Lil' Abner*. They call the process "shmooing." To mate, the shmoos fuse, of course.

Close-Up on Plantae

Plants. They're green, immobile, and showing off about making their own food. What else is there to know? Well, for many obsessive folk, there's no kingdom as diverse and fascinating and bizarre as the kingdom of the plants. What's so attractive about them? Is it the ovary hairs in citrus fruits? The orchids that look like the hind-parts of insects? Let's take a look.

Algae

Green algae are essentially like very basic water plants, and include *Chalmydomonas*, *Volvox*, and *Spirogyra*. They are much like plants, using similar chloroplasts, two types of chlorophyll, photosynthesizing, and storing starch. In fact, some are the closest relatives of land plants. So why are they not plants?

For one thing, they can exist as unicellular, colonial, or multicellular organisms, which is very unplantlike. A familiar multicellular form is what we call seaweed. One versatile little algae, *Volvox*, can do all three.

Algae exhibit alternation of generations, with the haploid form and the diploid form usually being about the same size. They require water to reproduce, having gametes with flagella, and we classify algae based on where the flagella emerges from the gamete. While some have placed algae into that grab-bag of a kingdom, Protista, doing so is controversial. And although some members—such as the green algae like the charophytes—are unquestionably the closest living relatives of land plants, even among the thousands of algal species, the relationships can be unclear. Brown algae (kelp) and yellow-brown algae, for example, may fall into the kingdom in progress, the Chromista.

Biohazard!

Don't be fooled by seaweed. You may think that seaweed, especially the large, green kind, is a marine plant. It is really algae and does not have the organs or organ systems that plants do. Yes, algae can be that big.

Mosses and Worts: Life Cycles

Liverworts, hornworts, and mosses (phyla Hepatophyta, Anthocerophyta, and Bryophyta, respectively) cross over into the "absolutely they're plants" zone. They also exhibit alternation of generations. This group belongs together for two reasons: they lack a vessel system for nutrient and water transport, and they do not make seeds.

With this group, you can see the green gametophyte, a haploid structure anchored by rhizoids (long tubular cells or cell filaments) rather than true roots. Gametes require water to reach the egg, which sits in *archegonium* (plural, archegonia), so these plants are usually closely tied to moist environments. The male gamete is produced in the *antheridium* (plural, antheridia).

def•i•ni•tion

The **archegonium** is the female reproductive organ in seedless plants and some conifers, a moist environment where sperm, produced in an **antheridium**, arrive for fertilization. The **sporangium** is the structure, at the tip of the resulting sporophyte in mosses, that contains haploid cells.

The resulting zygote grows into a sporophyte, still attached to the gametophyte and getting food from it. A *sporangium* (plural, sporangia) at the sporophyte tip contains haploid spores arising from meiosis. Once these are released, new green gametophytes will grow from them. So the sporophyte grows on the gametophyte and produces the haploid spores that will give rise to a new gametophyte.

Bio Basics

While most vascular plants you may recognize live on land, plants also can be aquatic organisms. The water lily is an example.

Most seedless vascular (with a vessel system) plants, like ferns, are homosporous, meaning that they make one type of spore that becomes a bisexual gameto-phyte. Heterosporous plants, primarily seed plants but also mosses, develop male and female spores. When released, these grow into either female or male gametophytes. Each spore first grows into a protonemata (singular, protonema), filaments one-cell thick that grow into the green, leafy-looking gametophyte that we see as "moss." The table straightens this out for you.

Gametophyte vs. Sporophyte

Type of plant	Features
Nonvascular plants (e.g., mosses)	Dominant, free-living gametophyte, smaller, dependent sporophyte. Usually homosporous, except mosses.
Vascular, seedless plants (e.g., ferns)	Reduced, free-living gametophyte, dominant sporophyte. Homosporous.
Seed plants (e.g., pines, daisies)	A lot like us—microscopic gametophytes, dominant sporophyte. Heterosporous.

Clothing Optional: Naked Seeds

Everything about mosses and worts starts to change when we get to ferns (Lycophyta = club mosses; Pterophyta = ferns). Now the sporophyte is dominant, so the green stuff you see on a fern is the diploid part. Underneath those feathery, ferny leaves are the sporangia, harboring the haploid spores. The spores grow into teensy bisexual game-tophytes. The male gamete still requires water to swim to and fertilize the egg lying sheltered in the archegonium. Although these plants do not form seeds, they are vascular and can grow tall against gravity.

Seed plants are divided into two groups: the gymnosperms (such as conifers and including the phyla/divisions Gingkophyta, Cycadophyta, Gnetophyta, and Conifero-phyta), with their "naked" seeds; and the flowering plants or angiosperms, which have seeds that "wear clothes."

Conifer seeds are "naked" in that they're not tucked away in an ovary. In this, they are closer in ways to the cell in the archegonium of a seedless plant like the fern. Also like the fern, the seed sits on leaves. In conifers, the leaves have taken on protective characteristics, like being extra tough and woody. We call these leaves grouped around the naked seeds "cones." As with pines today, these naked seeds closely protected in a cone may have arisen to withstand drier conditions.

Flowering Plants: Colorful Co-Evolution

Angiosperm seeds are "dressed" in fruits, which are mature ovaries harboring seeds. The day I learned that the little squirty things in an orange are called ovary hairs opened a new view for me on citrus fruits. Chapter 20 brings greater detail on seeds and fruit and the sex lives of plants.

Angiosperms fall into two groups, depending on whether or not their embryos put out one cotyledon (seed leaf) or two. If it's one, the plant is a monocot; if it's two, we used to call it a dicot, but now we call it a eudicot.

Close-Up on Animalia

This header is slightly misleading. We're going to spend several chapters on a real close-up on Animalia. Here, I give you two tables summarizing very briefly some of the vast diversity of the animal kingdom.

Invertebrates: Floppy but Fascinating

Invertebrate means "has no backbone." It's an enormous group.

A Selection of Invertebrates

Animal Phylum	Familiar Representatives	Features
Annelida	Earthworms	Segmented body, rudimentary kidneys
Arthropoda	Crabs, shrimp, spiders, insects, centipedes	Segmented exoskeleton, jointed appendanges
Calcarea	Sponges	Sessile, lack true tissues

continues

A Selection of Invertebrates (continued)

Animal Phylum	Familiar Representatives	Features
Cnidaria	Jellies, corals	Radial body plan, gastrovascular cavity, sting-y
Echinodermata	Sea stars	Bilateral larvae, five-part adults, internal skeleton
Mollusca	Clams, snails, squids	Soft bodied, nervous system, mantle for breathing and excretion, muscular "foot"
Nematoda	Hookworms, pinworms	Cuticle coating the body, some hermaphrodites
Platyhelminthes	Planarians, tapeworms	Bilateral symmetry, central nervous system, unsegmented

Vertebrates and Their Evolution

Then we have the vertebrates, animals with backbones. Confusingly, the Cephalo-chordata and the Urochordata don't have backbones, but developmentally they have vertebrate characteristics. What happens during juvenile development can be as informative to classification as the adult versions.

A Selection of Vertebrates

Phylum	Representatives	Features
Cephalochordata	Lancelet	Vertebrate "outgroup"
Urochordata	Tunicates	Larvae are "chordate-like"
Petromyzontida	Lampreys	Jawless fish
Chondrichthyes	Sharks, skates, rays, chimeras	Cartilaginous skeleton
Actinopterygii	Ray-finned fishes	Bony skeletons, fins are rayed
Actinistia	Coelacanth	Lobe-finned fish thought extinct, turned up in Indian Ocean

Phylum	Representatives	Features
Amphibia	Frogs, salamanders, newts, caecilians	Moist skin, usually an aquatic and land phase of life
Reptilia	Turtles, snakes, lizards, tuataras, birds	Amniotic egg, lungs
Mammalia	Monotremes, marsupials, eutherians	Fur, lactation

The Least You Need to Know

- Prokaryotes are distinguished as Archaea or Bacteria primarily based on nucleic acid sequencing.

- Of the eukaryotic kingdoms, Protista is a grab bag of widely different organisms.

- Fungi are distinguished by cell walls of chitin and external digestion and often eat dead stuff.

- Algae sit at the border between protists and plants, and green algae are the closest living relatives of true plants.

- Plants are seedless nonvascular (moss), seedless vascular (ferns), or seeded vascular (conifers and angiosperms).

- Animals, which are mobile and multicellular, are divided into vertebrate and invertebrate phyla.

Part 5

Form and Function (Plants and Animals)

In shaping organisms, nature selects features and forms that fit the organism's function in its environment. This powerful link between form and function has always informed our studies of how to categorize organisms and understand their inner and outer workings. From earliest development to adulthood, nature has programmed organisms to have specific forms, including configuration of the intricate network of body systems that keep us nourished and balanced even as the environment changes around us.

Chapter 20

Plants

In This Chapter

- ◆ The many roles of the meristem
- ◆ Plant tissues
- ◆ Roots, stems, and leaves
- ◆ Seeds and fruit

For land plants, form follows function. To live on land, immobile plants need ways to acquire water and nutrients from soil and CO_2 and light from above the soil. The forms they take enable them to seek light and to pull nutrients from many feet into the dark earth. Because plants can't walk, the forms they take must also enable them to adapt for immobile mating and reproduction.

It All Stems from the Meristem

Think about all the organs you have in your body. Now think about plants. They have three organs: roots, stems, and leaves. That was easy.

The stems and leaves form the shoot system, which starts with the shoot sending carbohydrates to the roots, which then have energy to send minerals and water to the shoots. Their growth is vegetative, or nonreproductive, growth.

Leaves are responsible for regulating water content, photosynthesis, and sex. Yes, sex. Those beautiful flowers are really leaves especially modified for sexual reproduction. Their growth is reproductive growth.

Like our organs, plant organs consist of tissues. Every plant tissue gets its start with the plant equivalent of stem cells, the meristem. And by "stem cells," I don't mean cells in the stem; I mean cells from which other cell types are derived.

Components of the Meristem

Plants start with two basic tissue progenitors in the meristem. From there, they can exhibit *primary growth*, which involves a set of three tissue systems and their related tissues; or they can exhibit *secondary growth*, which has some overlap with tissues from primary growth. In the beginning, primary or secondary, it all stems from the meristem.

def•i•ni•tion

Primary growth refers to tissues involved in lengthening the plant. Secondary growth, exhibited by seed plants, involves tissues necessary for widening the plant.

The meristem consists of permanently embryonic tissues of two types. At the root tips, where they get pointy, we have the apical meristem, which also extends to shoots and axillary buds. Their job is primary growth or growing things longer, the roots pushing downward and the shoots pushing upward. Adding thickness, or secondary growth, is the job of the lateral meristem.

From the apical meristem, three types of primary meristem arise that will differentiate into three types of plant tissue systems. The protoderm around the outer stem will become the plant's dermal tissues, while just underneath lies the procambium, which will differentiate into the plant's vascular tissues. Finally, there is the ground meristem, which gives rise to the plant's ground tissues.

Biohazard!

Crossover confusion may assail you here. Note that in some cases, the same tissue type (e.g., phloem or xylem) can be produced from primary meristem tissue systems or from lateral meristem. We say that these tissues can be primary or secondary in origin.

The lateral meristem, which contributes to widening the plant, occurs as two types, the vascular cambium and the cork cambium. A widening plant has to have tissues in its widening parts just as it does in its lengthening parts, so some tissues that arise from primary meristem origins can also arise from lateral meristem origins, including xylem and phloem.

The cork cambium is a cylinder of cells that gives rise to cork, which replaces the outer layer of plant tissues as the plant widens. You may have seen these layers in a slice of tree trunk. The cork cambium also makes the periderm, which together with some phloem makes that rough outer covering of the tree we know as bark.

The vascular cambium gives rise to the vascular tissues that populate the secondary growth of the plant, the xylem and phloem, which also can arise from primary meristem tissue systems. When xylem from secondary growth becomes abundant enough, the plant has become woody, or formed wood.

Plants Have Tissues, Too

Each of the plant organs arises from three types of tissue systems: the dermal, vascular, and ground tissue systems.

Dermal Tissue System

Our dermis is our skin. The plant's dermal tissue system also produces its outer covering. As with us, the dermal tissue is a general defense against exposure to germs or internal damage. It may include the epidermis, which might have a cuticle, a layer of protective wax, on it. Plants that have gone woody (have a lot of xylem) convert to periderm, which is an aggregate of a few different kinds of tissue and the result of secondary growth.

While each organ has dermis, the dermis does different things for each organ. In leaves, the dermis might secrete oils that offend predators, while root dermis might have extensions that increase nutrient absorption.

Vascular Tissue System

The two tissues in the vascular tissue system are xylem and phloem. In Chapter 3, we encountered the concepts of adhesion and cohesion of water molecules. Water (and things dissolved in it) can be pulled upward from the roots by evaporation through the xylem, pulling lower water molecules higher through adhesion. At the same time, the water molecules can cohere to the walls of the tube.

For sugar transport, plants use phloem, moving sugar usually from top (leaves) to bottom (roots). In plants, xylem and phloem together with procambium and other tissues make up the stele.

Ground Tissue System

The nondermal, nonvascular tissue is the ground tissue system. If it's inside the vascular tissue, it is pith; if it's outside, it is cortex. This tissue serves many purposes, including storage and support.

The three primary tissues of the ground tissue system are *parenchyma, collenchyma,* and *sclerenchyma*. Parenchyma, the most common, occurs all over the plant, in the roots, stems, and leaves and amidst the vascular tissues, primarily as a soft, spongy tissue. Collenchyma, a stretchy tissue, has thickened cell walls and provides support. Sclerenchyma, distinguished by being dead at maturity, consists of cells with thick walls that also provide strength, support, and sometimes storage. It also is the tough coat that surrounds seeds and forms nutshells.

> **Bio Bits**
>
> The best-known form of sclerenchyma is probably the fibrous and occasionally infamous product known as hemp.

Roots: Specialized for Uptake

Root growth occurs in three zones. The zone of cell division, near the protective root cap, is the site of mitosis. Root lengthening takes place along the central part of the root, in the aptly named zone of elongation. Maturation of root cells takes place in the part of the root nearest the shoot, in the zone of differentiation. These zones don't have little fences dividing them, and there is some overlap.

Nutrient Uptake

Gymnosperms and eudicots (like pecan trees) have tap roots, the root that emerges first from the embryo. These are great for reaching deep for water. From this central root, lateral roots poke out. If you've crunched a carrot, you've crunched the tap root, where much of the plant's sugar is stored.

Monocot roots differ. Think grasses. Short, furry roots growing from the stem, forming a mat. You can't roll out a lawn of pecan trees, but you can roll out mats of St. Augustine grass. That's because the tap root regressed. These mats of roots serve in erosion control.

At the root tips are a bunch of root hairs, which are cell extensions, not to be confused with lateral roots. Like the folds of the small intestine, these root hairs increase the absorption area for the plant.

Symbiosis: Living Together for Life

Plants also form friendly partnerships with other organisms. Perhaps the best known is the pairing of legumes, like soybeans, and friendly bacteria that fix nitrogen. Living things need nitrogen to make most of the Big 4 biomolecules (see Chapter 4), but most plants and animals can't just take N_2 and use it; they need it as ammonia. These plants use the bacteria to grab N_2 and build accessible ammonia, which the plant then uses. The bacteria collect in a structure called a nodule on the legume root, where it finds fairly safe shelter.

Another relationship is the partnership, or mycorrhizal association (*myco* = fungal), of plants and fungi, in which the fungus sets up shop on the plant's roots. The plant helps out the fungus with carbs, while the fungus helps the plant absorb soil nutrients.

Stems: Supporting and Transporting

A node is where the leaf attaches, and the stretches in between these nodes are, naturally, internodes. A branch is a budding off where the leaf and stem form an angle, from which extends the axillary bud. Lengthening happens at the plant tip, where the apical bud is. Because of its elevated position, the apical bud gets most resources early on, reaching the light first.

The stem's role is to support the plant, pushing toward the light. It also contains the vasculature. Sometimes, a stem doesn't look like a stem. The potato, for example, is a tuber, a shoot that grows under the soil surface and stores food, which is why it's so starchy.

Bio Bits
If you pinch off an apical bud, axillary buds will get redirected nutrients and begin to flourish. This is why people like my mother go around pinching things off of hedges—it makes the shrubs grow out fluffier around the sides and not at the top.

Leaves: Food and Regulation

The leaves are the solar panels of the plant, the place where most photosynthesis takes place. Through the specialized stomata, leaves also control how much water a plant retains or releases. Plant classification relies largely on leaf features, including shape, arrangement, and how the veins branch.

A leaf consists of the blade, often that familiar teardrop shape, and a stalk, called the petiole, where it attaches to the plant. The arrangement of the blades can be as simple as a single blade attached to the stem, or as complicated as subleaves occupying a single petiole, called a compound leaf. Even more complex are the double-compound leaves, with sub-subleaves making up subleaves that make up a larger leaf. The veins are, like our own veins, the vessels through which water and nutrients reach these critical structures. Remember that leaves must have water to initiate the process of photosynthesis.

The arrangement of leaves also offers layers of classification involving terms like pinnate and palmitate that we won't detail here. Other leaf features that serve to classify plants include their shape, their margins or edges (serrated? smooth?), and, in the case of compound leaves, how they're arranged on the stem.

Internal Structure

While leaves offer diversity on the outside, on the inside they have much in common because their functions for the plant, regardless of species, stay the same. On the outside of the leaf is its epidermis. This waxy skin, like yours, helps the leaf in several ways, including gas exchange, protection, and water regulation. One of the key features of the epidermis is the stomata, the pores that regulate water loss. They consist of two guard cells with a gap between them that can open and close depending on the plant's water needs.

Deeper down in the leaf is the mesophyll, where photosynthesis takes place. While the epidermis is fairly colorless, having no chlorphyll, the mesophyll is chock full of pigment and gives plants their green color. In many plants, the mesophyll has two layers, the upper palisade layer with lots of chloroplasts, and a spongy lower layer.

> **Bio Bits**
>
> Carnivorous plants are plants that eat animals. One kind, the pitcher plant, is so specialized that its "pitcher" consists of waxy leaves angled so that any hapless insect landing on the edge finds itself on a deadly superslide to the plant's interior swimming pool of digestive enzymes.

Running throughout the leaf's spongy mesophyll layer is its vasculature, the veins. They consist of the now familiar xylem and phloem. The special pattern that a leaf's veins make is called venation.

Special Jobs for Leaves: More Than Photosynthesis

In addition to serving as the primary seats of photosynthesis, leaves aid in sex. The attraction can be a simple matter of bright color combined with an attractive scent thanks to nectar, or it can be more complicated, such as the orchids that mimic a female wasp's hindquarters to attract the male as a pollen-transfer mechanism.

Seeds and Fruit

We met sporophytes and sporangia when we talked about nonvascular and seedless plants in Chapter 19. Seed plants have not given up on these structures, but now they've gone micro and mega. In the parent seed plant, the sporophyte is home to a megasporangium and megaspore that will become the female gametophyte or a microsporangium that produces the microspores that will become the male gametophyte. In the sporophyte, a protective tissue layer called the integument protects the megasporangium.

How the plant brings the pollen to the egg depends on the plant. In gymnosperms (e.g., conifers), the cone can be destined to produce either a male or a female gametophyte; conifers often can make both. The woody scales of the cone contain and protect the sporangia, where meiosis produces haploid cells. In female cones, one of these cells is the megaspore that gives rise to the female gametophyte. This gametophyte makes two or three archegonia, each containing an egg.

 Bio Basics _____

Note that now with seed plants, the sporophyte bears the sporangia. Note also that as we've gone from mosses to ferns to seed plants, the sporophyte has become dominant and the gametophyte has become almost microscopic.

The female cone doesn't even get into this process until a pollen grain arrives. Pollen is produced in the male cones: the microspores in the microsporangia produce the male gametophyte, or pollen, which has two sperm nuclei. The wind carries the pollen, buoyed by an attached air sac, to the female cone, and it gets stuck in a pore called the micropyle, which leads to the megasporangium. While this process is called pollination, it is not officially fertilization yet.

For fertilization to happen, the pollen extends a pollen tube into the micropyle and to the megasporangium. This extension triggers the megasporangium to produce the female gametophyte and the archegonia with their eggs. Ultimately, one of the sperm nuclei will fuse with one of the eggs to achieve fertilization. The delay between pollination and fertilization can be years. You'll notice the wind is the agent that gets the pollen where it's going.

Angiosperms, the flowering plants, have a slightly different approach. The male plants (or their male-producing parts) have stamens, consisting of filaments with anthers at the tips, where the plant makes microspores. If you peek inside a flower and see fuzzy yellow pollen on its structures, you're looking at its anthers. Microspores are destined to become pollen grains that, as in conifers, travel as paired nuclei via insect or wind and arrive at the female to get stuck in a tube. In angiosperms, the tube is called a stigma. The pollen then forms a pollen tube that leads into the ovary.

The ovary is a vase-shaped structure that is part of the pistil or female parts of the plant. In the pistil, the stigma leads to a style, a tube that stretches into the ovary. In the ovary lie the customary ovules with megasporangia that contain megasporocytes that produce megaspores by meiosis. One of these will become the gametophyte. One pollen nucleus that travels down the pollen tube fuses with it to form the zygote. The other pollen nucleus fuses with the central cell, a double-nucleus cell also in the ovule, and forms a triploid tissue. This triploid tissue is the seed's endosperm, which serves as the seed's food supply through development.

Bio Basics

Fruits are often not what we think. For example, grains like barley and rice are fruits. We think of the red part of the strawberry as being fruit, but the real fruits there are the little dried ovaries all over the outside.

The seeds of angiosperms are not naked; instead, they sport a diploid seed coat. In addition, some seeds don even heavier attire in the form of the flesh of fruits or the tough coating of the nut. Either way, the tasty contents attract organisms that can help disperse the seeds. Something to ponder the next time you eat fruit: that's a mature ovary you're eating there.

The Least You Need to Know

◆ Roots, stems, and leaves, respectively achieve nutrient uptake, growth, and photosynthesis/reproduction for the plant.

◆ Primary meristem produces primary (upward) growth and lateral meristem (vascular and cork cambium) secondary growth.

◆ Protoderm, procambium, and ground meristem of primary meristem produce the dermal, vascular, and ground tissue systems.

◆ Dermal gives rise to epidermis, vascular to xylem and phloem, and ground to collenchyma, parenchyma, and sclerenchyma.

◆ At pollination, pollen extends a pollen tube into the ovule, while fusion of a pollen nucleus and egg is fertilization.

◆ In angiosperms, the second pollen nucleus fuses with a diploid female cell to form endosperm for the seed.

Chapter 21

Animals, the Basic Plan

In This Chapter

- ◆ The importance of the body cavity
- ◆ Symmetry in form
- ◆ The early three: ectoderm, mesoderm, and endoderm
- ◆ Two basic cell types: epithelium and mesenchyme
- ◆ Overview of organs and organ systems

In keeping with our theme—and nature's theme—of the relationship between form and function, here we bring you the ways animal forms are shaped to execute function. Forms are no accident. They arise from a carefully orchestrated and selected process of development and cellular decisions that all begins with a single cell.

The Simplest Plan

We tend to think of animals as four-legged creatures, possibly with fur, but as the animal kingdom survey in Chapter 19 showed us, they also cover everything from sponges to spiders. The simplest animals consist of collections of specialized cells that are not quite true tissues but are distinguishable as different cell types.

Bio Basics _____

While sponges don't wear clothes and flip krabby patties for a living, they're still a diverse bunch. Some are carnivores. While some can reproduce by budding, many are hermaphrodites. Although they're known for immobility, a few species can scoot very slowly across the ocean floor.

Until recently, sponges were considered to be the most basic animals. As adults, they're usually *sessile*, meaning stuck where they are, immobile. Their mobile stage is a brief period when they are larvae. At one time, sponges ruled the world, dominating the earth's shallow seas several hundred million years ago. Had a human suddenly shown up, it would have been an eerie place with just the sponges, the seas, and some invisible microbes sloshing around.

> **Bio Bits**
>
> The only known living placozoan is *Trichoplax adhaerens*. It first came to notice as a spreading growth on the side of an aquarium in Austria in the nineteenth century.

We've now discovered an even simpler form of animal, the placozoan. This placozoan qualifies as an animal primarily because it is multicellular, doesn't photosynthesize, and doesn't externally digest its food. It looks like an amoeba, but it has three kinds of basic tissues and a small internal cavity. It moves using cilia and is only a few cells thick, barely visible to the naked eye. Yet, it is the most ancient animal lineage known. It even may have the chemical basis of what eventually evolved into the animal nervous system.

A Body Cavity

The body cavity, or coelom, is a fluid-filled space that separates the inside tissues from the outside tissues. Some animals, such as flatworms (e.g., Planaria) lack this internal cavity, but had one at one time in their evolutionary history. These animals are known as acoelomates. Insects have a coelom that serves as an open circulatory system called a *hemocoel* (*hemo* = blood).

The gut that this cavity separates from the outer tissues arises in one of two ways. Animals, including us, start out as a ball of undifferentiated cells. The first distinguishable body structure to form is the body cavity. It starts in protostomes (*proto* = first; *stoma* = mouth) with cells that form a tube by migrating inward where the mouth will be (*protostome* = mouth first). It starts in deuterostomes with cells migrating inward where the anus will be (*deuterostome* = mouth second). We and many other animals, including starfish and all vertebrates, are deuterostomes that form the anal opening first and the mouth second. Protostomes include earthworms and arthropods. Regardless of where it starts, the end result is a primitive gut called the archenteron. Surrounding it is the fluid-filled coelom.

Bio Basics _____

Animals were once classified based on having a coelom or a pseudocoelom, but this classification is not entirely accurate. Even though the "pseudo" implies that pseudo-coelomates like roundworms lack a true body cavity, that's not the case. They have a coelom, it's just derived from two different tissues, rather than one. All pseudocoelomates are protostomes.

The fluid-filled cavity provides space and protection for organs, and many animals have evolved internal organs that occupy the opening. Because they're not stuck to the outside tissues, these organs can move on their own without affecting outer movement. Your stomach can squeeze tight in the contractions of digestion without your even noticing it, thanks to this separation. The coelom also serves as a sort of fluid-pressure, or hydrostatic, skeleton for some organisms lacking other internal skeletal structure, such as earthworms.

Bilateral Symmetry: Right and Left

All but the simplest animals have symmetry. Using a food analogy, animals are blobs, submarine sandwiches, or pies. Sponges don't have symmetry (same measurements): they are asymmetrical blobs. Cut into a sponge on any axis, and you'll end up with two parts of the sponge that don't exactly match. Radial animals are like pies—cut them along any axis, just as you would a pie, and you'll get two equal halves. Jellyfish (in the phylum Cnidaria) exhibit radial symmetry.

Then there are the bilateria, which comprise most animals, including us. They're the sub sandwiches. Cut them along the vertical axis, and you get two equal halves. But if you tried to cut them like you do a pie, on any other axis, the parts would not be identical.

Germ Layers

One chief characteristic of animals is their multicellularity. To be multicellular requires having cells with distinct duties. Some cells might work together in absorption, while another group might work together in defense. These cells working together as a unit form a tissue. But before these cell lines became tissues, they were all essentially exactly the same, all sprung from the one cell that formed when sperm and egg met. Something happened along the way to make one set of cells different from another. That something is called *differentiation*.

def•i•ni•tion

Differentiation is the process by which a generalized cell type, such as a stem cell, becomes more specialized. A differentiated cell will usually express only those genes necessary for it to function as its specific cell type. Differentiated cells usually cannot revert to their initial, more general condition.

The first step to differentiation is gastrulation. In this process, an embryo goes from being a ball of similar cells, one fairly indistinguishable from another, to forming different cell types. Gastrulation involves changes in cell motility, cell shape, and cell adhesion that allow changes in how the cells behave. It ends with the formation of the three basic tissue types present in most animals: ectoderm, mesoderm, and endoderm. These tissues are also called the germ layers because they form the seed or germ for the other tissues that differentiate from them. (We discuss gastrulation in more detail in Chapter 24.)

While most animals form these three layers and are known as triploblasts, a few groups of animals form only two layers, the ectoderm and the endoderm, and are known as diploblasts. This group of animals, which includes corals (yes, corals are animals) and jellies, also happen to be prone to radial symmetry. All bilateral animals are triploblasts.

Endoderm (*endo* = inner) is the tissue on the outside after gastrulation, mesoderm (*meso* = middle) is the tissue in the middle, and ectoderm (*ecto* = outer) is the outside tissue layer. These three germ layers are always depicted in diagrams as having characteristic colors: ectoderm and tissues arising from it are given in blue, mesoderm in red, and endoderm in yellow.

Ectoderm Gives Rise to ...

Organs and tissues that arise from ectoderm include your cornea and lens of your eye, your nervous system (yep, those early cells were on the outside before they migrated inward and became nervous system), your tooth enamel, and, of course, your epidermis (skin).

Biohazard!

What you read here refers primarily to vertebrates. Also, don't be misled by the prefixes. Sometimes, organs or tissues you might consider internal arise from ectoderm, including the lining of your mouth. Yes, that's inside, but in the beginning, it was on the outside. The cells whose descendents were destined to line your mouth migrated very early on from the outside in.

Mesoderm Gives Rise to ...

This one's a bit tricky to intuit. What's in the "middle" of your inside and outside? Well, among the organs and tissues arising from that middle germ layer are your bones and muscles, your circulatory system, the lining of that body cavity that formed early on, your urinary system, and most of your reproductive system.

Endoderm Gives Rise to ...

What's left? The inner layer of cells that forms during gastrulation eventually differentiates into several of the innermost tissues of your body. These include the lining of your digestive tract and respiratory system, your liver and pancreas, your thyroid, and the lining of your urinary system.

Tissues

There are many different kinds of tissues that form the complex organism that is you, but in the end, most of the cells that make them up can be categorized into two types.

Epithelium and Mesenchyme

It may be that an important moment of your life is gastrulation, when the big process of germ layer formation goes down. But there are many other smaller, Very Important Moments during development, often involving a shift from one general cell type to another. These two cell types are epithelium and mesenchyme, and they each have different significance in development.

When cells stick together in sheets, they form epithelium. They stick together because they express adhesion proteins on their surfaces that "hook" the cells together. These sheets of cells form, for example, the lining of your gut or lungs. They can adhere so tightly to each other that a tissue consisting of them can be watertight.

As you see in more detail in Chapter 24, nothing in embryonic development stays the same for long. Cells that hung together in sheets may suddenly separate and start shifting around, changing their locations and beginning to become a new cell type. These cells have shifted from being epithelial cells to being mesenchymal cells, or mesenchyme. Mesenchymal cells are loose, no longer adhered to each other because they've stopped expressing the adhesion proteins that hooked them to other cells.

This ability to release from the epithelial sheet and change location, to experience a new environment and become a new type of cell, is critical to appropriate embryonic development. A cell that doesn't successfully make this shift when the time is right can derail the entire process. If it's not in the right place at the right time, some step in development will likely go awry.

Teamwork: Cells That Work Together

As you can see, successful development of an organism requires a carefully timed, precise ballet of cellular movement and change. How do cells "know" when it's time to shift to being mesenchyme or epithelium, or to become endoderm or ectoderm, or to differentiate into urethra? As with any other form of communication at this level, the cells "talk" to each other via molecular interactions.

They may send chemical messages across the embryo that lead to changes in gene expression. They may alter gene expression in response to changes in how crowded the environment is or other signals about their location. And when cells destined to become a specific tissue all receive their appropriate signals and respond in the right way by expressing the right genes, they've become team players that contribute to the overall function of the organism.

Organs and Organ Systems: Preview

When it all works out like it should, cells assort themselves in sheets or loose aggregations, all expressing the right genes for their location and duties, forming tissues that create organs that make up organ systems. The following table shows some of the organ systems of vertebrates and their organs, the components of the complete, functional animal body plan.

Some Vertebrate Organ Systems and Their Organs

System	Organs
Circulatory system	Heart, blood, vessels
Digestive system	Salivary glands, esophagus, stomach, liver, gallbladder, pancreas, intestines, rectum, anus
Endocrine system	Hypothalamus, pituitary, gonads, adrenals, pineal gland, thyroid; also includes bones, kidneys, even heart
Integumentary system	Skin, hair, nails
Lymphatic system	Lymph fluid, nodes, vessels, thymus, spleen
Muscular system	Muscles
Nervous system	Brain, spinal cord, nerves
Urinary system	Kidneys, ureters, bladder, urethra

The Least You Need to Know

♦ The simplest animals have tissues and a coelom, a body cavity for holding organs or serving as a hydrostatic skeleton.

♦ Animals that begin coelom formation at the mouth are protosomes, while deuterostomes begin the coelom at the anus.

♦ Animals may have no symmetry or exhibit radial (like a pie) or bilateral symmetry.

♦ Most animals are triploblasts that start with three germ layers: ectoderm, mesoderm, and endoderm.

♦ Germ layer formation is a process called gastrulation, and each layer is predestined to develop into specific tissues.

♦ Cells hooked together in sheets are called epithelium, while those that are loose are mesenchyme.

Chapter 22

Systems 1: Homeostasis and Materials Exchange

In This Chapter

- Nutrition: we are what we eat
- The animal digestive system
- Maintaining fluid balance
- Circulation and gas exchange

We take in foods and liquids and get rid of the waste. We inhale and exhale gases, but what happens in between? Our bodies take what they need to maintain balance among all of the complex processes that keep us running smoothly. We use what we need and get rid of what we don't. That is, if everything is working like it should.

Nutrition and Digestion

We eat because we need atoms and molecules to build more atoms and molecules. As you read this, your body is burning through a million little physiological processes, using nutrients, breaking up sugars, releasing

def•i•ni•tion

Homeostasis is the maintenance of balance in the inner environment; in other words, keeping things running smoothly in you.

calcium, pumping sodium and potassium, breaking up ATP, building proteins, and on and on. It can't do this unless you give it what it needs for each of these many molecular pathways that keep you going. It's one way that your body maintains *homeostasis*, by replacing what it uses.

Food and Nutrition Guides

Things have changed since I was in elementary school back in the dark ages. I know it's practically a dull proverb, but things really were a lot simpler then. We learned about nutrition as the four food groups: starch, dairy, vegetables, and meat. So simple, so basic. The four together made what we called a "square meal."

Today, meals are pyramid shaped. The powers that be when it comes to nutrition have reshaped the square into a complex pyramid that shows in wedges the relative amounts of different kinds of food you should eat. The widest swath is the carbs—bread, cereal, rice. This idea that plant-based foods form the basis of our nutritive needs is a reflection of something you learn later: that plants themselves form the basis of an ecoystem's nutritive needs.

Someday, this pyramid may go the way of our square meals from the '60s and '70s, and your children or their children will be learning about some other geometrically oriented nutrition program, perhaps a dodecahedron. The bottom line when it comes to nutrition? Unless you have some special dietary requirement, simply consume all things in moderation, and get daily exercise. It really is that simple. No geometry required. Now I will duck as incensed nutritionists start throwing raw organic fruits and vegetables at my head.

From the Entrance to the Exit

Whatever you consume, your body will try to absorb the important parts. Sure, it also can absorb some things we don't need, like the ethanol in alcoholic drinks, but most things we eat have some kind of nutritive value that your entire digestive system is designed to grab and use.

The "use" part of your digestive system starts as soon as you put something in your mouth. Our teeth start the breakdown process by literally chewing the food into little bits. You also have enzymes in your saliva specifically designed to attack and break up starch while also—and pardon the grossness—getting your food lubricated for its trip down the esophagus into your stomach. On its way there, the tongue pushes the food through the pharynx and then into the esophagus. Because nature, in her infinite wisdom, branches the tube to the lungs (the trachea) and the tube to the stomach (the esophagus) from the pharynx, your body protects you from choking as you swallow by covering the trachea with the epiglottis, a flap of cartilage. It's just a fun irony that the two processes required for you to live—breathing and eating—have this little side-by-side glitch.

> **Bio Bits**
>
> Your digestive system can be as long as 9 meters, or over 30 feet.

Your esophagus has smooth muscle all along it that contracts in waves called peristalsis to push the food down into your stomach. A tight ring of muscle called a sphincter (you've got more than one of these in your body) lies at the end of the esophagus. The muscle contractions of the esophagus push the food through this muscle and into your stomach.

Now the real breakdown takes place. Your stomach is not a bucket of acid that atomizes food. It's a complex place involving interactions of mucus, enzymes, and acid. The acid doesn't do so much to break down your food as it provides the right pH for the digestive enzymes to break down your food. If you think back to lysosomes (Chapter 5), they maintained a very low pH so that the hydrolases they contained could break down macromolecules. Your stomach operates in a similar way. The acid is also a pretty uncomfortable place for some of the microbes that may have hitchhiked this far into your digestive system.

The enzymes that break down your food include pepsin, which breaks down proteins, and enzymes specialized to break down some fat and carbs. The mucus protects your stomach lining from this rather unpleasant environment. With the cells under mucosal protection, there is no absorption taking place here. That starts in the small intestine, where your food goes from your stomach, exiting through another sphincter, the pyloric sphincter.

Your small intestine is beautifully designed to absorb nutrients. It's lined with specialized cells that have two distinguishable sides, the apical face and the basal face. The basal face is turned inward, away from food contact but in contact with your circulation. It is specially designed to interact with your circulation, transferring nutrients to be sent via these fluid pathways to your body's cells.

Meanwhile, the apical surface, facing into the intestine, is especially designed to maximize the amount of surface available for nutrient absorption as the food moves past. It's got fingerlike projections called villi that, in turn, also have fingerlike projections called microvilli. To see how this increases nutrient absorption, make a fist with your hand. Note how much skin you've got available to interact with the environment. Now open your hand and spread out your fingers. You've just increased the available area significantly. Now imagine that your fingers sprout little tiny fingers all over. Even more area. That's what the lining of your small intestine is like. These fingerlike extensions wave around in the open space of the intestine, called the *lumen*, grabbing any nutrients that pass by.

def•i•ni•tion

A **lumen** is the open space within a tubelike body structure, including gut, bladder, and esophagus.

The 20-foot-long small intestine consists of three parts, the *duodenum*, *jejunum*, and *ileum*. The duodenum continues the process of digestion, while the other two components do most of the absorption of nutrients, including water.

The final part of the journey is through the four-part *colon*, which takes up any last-minute water and salt before any remainder moves on to be eliminated through the rectum and anus as solid waste. We also have bacteria living in this part of our digestive system, which work with us to produce some vitamins that our cells need (such as some B vitamins and vitamin K) while we give them a nice, dark place to live filled with, um, passing nutrients.

Bio Basics

Digestive system summary: mouth (some digestion)→pharynx→esophagus (peristalsis)→ stomach (digestion)→duodenum (more digestion, pancreas helps)→jejunum and ileum (absorption)→colon (water absorption, friendly bacteria)→out (toilet tissue).

You are likely familiar with the ultimate outcome of this process, so I don't feel like we need more detail on that. As one of my children's books informs us, *Everyone Poops* (Kane Miller, 2001).

Digestive Paths Compared: You vs. the Sponge

What I've just described above is how omnivores like us "do" digestion. The basic setup is similar for many animals: a tube lined with cells specifically designed for maximum nutrient absorption. Even the lowly sponge, which doesn't even have true tissues, has an analogous construct. It doesn't have a mouth, but it uses pores to control water flowing through a cavity that runs in canals through its body. Along the walls of this cavity are cells that can snag food—mostly bacteria—from the water as it flows by. The entryway to the sponge's "digestive system" consists of cells called ostia. Flagella on the inner cells beat the water through in one direction, and the flow is ejected through exit pore cells called oscula.

It's not quite as complex as our system, but the principle is the same: movement of nutrients past cells specialized to pick them up for the organism's use.

Specialized Digestive Structures

Your pancreas has many jobs, but digestively speaking, its role is to produce some of the enzymes that continue food breakdown in the duodenum and to shift the pH of its contents from burning stomach acid to about 8. Most protein digestion occurs thanks to these pancreatic enzymes, trypsin and chymotrypsin. While the pancreas is probably best known for its endocrine (hormone-related) activities, this delivery of protein-breaking enzymes is one of its *exocrine* functions. It also produces enzymes for breaking down fat and starch in the gut, and pumps bicarbonate into the duodenum to buffer stomach acid pH.

The liver's big role in digestion is to produce bile. Bile is stored up in the gall bladder and used in the duodenum to help emulsify, or break up, fat. Fat, being hydrophobic, requires this special processing so that its nutritive value can be exploited to the utmost.

def•i•ni•tion

An **exocrine** gland secretes its molecules into ducts, rather than into the blood. The pancreas is an endocrine gland that secretes the hormone insulin into the blood, where it facilitates sugar transport into cells. But it also secretes enzymes into pancreatic ducts and is an exocrine gland.

Does Your Brain Listen to Your Stomach?

Do you ever feel like anxiety makes you have a knot in your gut? Have you sometimes felt that your stomach knows there's food around before your brain does? The "little brain" in your gut could be the nerve bundle responsible for your sensations. This enteric nervous system primarily has the duty of managing all of the processes of digestion, governing everything from when you swallow to when you … eliminate.

Like the big, bihemispheric bundle of nerves in your head, this little brain also consists of nerves and uses chemical signaling as an independent, near-sentient bundle of nerves in your gut. In fact, the two brains likely communicate with each other, and what affects one is probably going to affect the other. Researchers are discovering more about the link between these two "brains," including why "mental" stress produces gastrointestinal symptoms and antidepressants have gastrointestinal side effects.

Urinary and Excretory Systems

We just spent some time on the less-than-savory topic of solid waste. You may have noticed that solid waste isn't all that you eliminate. In fact, excretion technically refers more to what we urinate or exhale than to offloading solid waste. We spend time getting rid of liquid wastes, as well. This chapter is about materials exchange and maintaining a balance. When you take in food, you take in materials. Your colon aids in fluid balance, but the primary players in maintaining appropriate fluid levels for your body are the kidneys. They are critical to many aspects of staying alive, including maintaining blood pressure.

Biohazard!

Use the term excretion correctly. It refers to the release of metabolic waste products that cross a plasma membrane.

Your urinary system consists of your kidneys, which dump fluid into the ureters, tubes that drain into your bladder. From your bladder, the fluid exits through the urethra and out into the world. The fancy word for this is micturition. Most people just call it "peeing."

Fluid Balance: A Delicate Do-Si-Do

Maintaining the balance of fluids in your body is a highly tuned process. You're aware of it because you micturate, but you may not be aware of all the work that goes into filling your bladder or of all the other critical factors involved. Whether to retain or lose water is a decision that your body makes many times a day.

Animals and Fluid Balance

A single-celled organism maintains fluid balance through the passive process of osmosis, the movement of equilibrium-seeking water molecules from high to low concentration. Animals that live in water must work to maintain a balance of fluid and salts. Those living in saltwater, for example, have special mechanisms to maintain their internal water concentrations against the relatively low-water/high-salt seawater they're living in.

Animals that have packed up and moved to land have a different set of issues to manage. Insects have structures called Malphigian tubules that help retain body fluids while getting rid of wastes. Earthworms have a set of proto-kidney-like structures called nephridia that serve much the same purpose: filtering out wastes and retaining fluids as needed to maintain the proper water balance for the organism.

Mammals and Fluid Balance

Mammals (and all other vertebrates) use kidneys for water balance. This pair of bean-shaped organs works primarily to maintain fluid balance and blood pressure, and excretion is almost a side effect of their work.

Kidneys: Structure and Function

Kidneys consist of an outer part called the cortex, where most of the filtering action is, and an inner part called the medulla. All along the cortex are millions of complex little structures called nephrons, which most books will importantly tell you are "the functional units of the kidney." And indeed, it's true. Each nephron begins at the Bowman's capsule, a capsule around a wad of capillaries known as the glomerulus. (The kidney has some of the most tongue-tying terms in all biology.) All along the pathway described below, blood vessels are available to take up anything the kidney decides should be filtered back to the body.

The filtrate (liquid with stuff in it from the blood) moves from these capillaries into the Bowman's capsule and then flows into the proximal convoluted tubule. As its name implies, it's proximal (nearest the entry), it's twisty (convoluted), and it's a tube (tubule). Here, some important molecules (like ions and glucose) that the body might still need get filtered back into the blood through an active (energy-requiring) process. The filtrate then moves into the descending limb of the loop of Henle (named after its discoverer, Friedrich Gustav Jakob Henle). Once again, the name says it all. It's a tube that forms a loop, and this part of it drops down. Here, water gets shifted back to the organism and the filtrate becomes more concentrated.

> **Bio Bits**
>
> Because much of the water return to the body happens in the loop of Henle, longer loops mean more opportunity to return water to the body. Mammals that live in the desert are known for their long loops of Henle.

Rounding the loop of Henle, the filtrate moves into the ascending limb of the loop of Henle, where solutes—more ions and stuff—get ditched. Now it's time for a trip to the distal convoluted tubule (that's the twisty tube near the exit), the nephron's last chance to pick up any remaining wastes before the party's over. The next stop is the collecting duct, the final opportunity to dump some water back to the organism before the remaining filtrate flows into a large collecting vat called the renal pelvis, which funnels all of the filtrate from all of the nephrons of the kidney into the ureters and finally to the bladder.

Bio Basics

The kidney is not the only organ with a cortex and a medulla. These terms are used generally to indicate an organ's outer (cortical) and inner (medullary) parts.

Each nephron lies in both the cortex and the medulla of the kidney, and as the filtrate travels its route, it encounters a gradient of solute concentrations. The convoluted tubules lie in the cortex of the kidney, where the outer solute concentration is low. As filtrate moves down to the loop of Henle, this part of the nephron lies in the medulla, where solute concentrations increase. At the bottom of the loop, the outer solute concentrations are at their highest, leading water to exit the loop. As filtrate moves through the ascending loop of Henle, back up toward the cortex, the solute concentration decreases again, and solutes in the filtrate may exit here.

Kidneys and Processes: Why You Need at Least One

The filtration of wastes and rescue of necessary ions, glucose, and nutrients are just some of the work of the kidney. That last little bit of the filtrate pathway, the collecting duct, also contributes to some other mission-critical aspects of being a vertebrate.

Bio Basics

Urinary system summary: blood to nephron (million of them/kidney)→glomerulus→ Bowman's capsule→proximal convoluted tubule (in cortex, low-solute environment, solutes out)→descending loop of Henle (into medulla, high-solute environment, water out)→ascending loop of Henle (toward cortex, low-solute environment, solutes out)→distal convoluted tubule (cortex, last-minute waste pickup)→collecting duct (water filtered back to body if ADH present)→ureters→bladder→urethra→out (micturition).

When you're thirsty, your brain sends a chemical messenger to this part of the nephron, a hormone called antidiuretic hormone (ADH). A diuretic makes you lose water, so an antidiuretic is a signal to retain water. If ADH is present, it makes the walls of the collecting duct permeable to water, and more water moves from the filtrate back to you. If no ADH is present, the water stays in and moves to your bladder.

In addition to maintaining this fluid homeostasis, the kidney also oversees your blood volume, which in turn plays a role in blood pressure. If blood volume drops, which also means a blood pressure drop, a signaling molecule called angiotensin causes the adrenal gland to release another signaling molecule called aldosterone. Aldosterone acts on the cells of the distal convoluted tubule to make you retain more water and sodium. At the same time, angiotensin acts on your blood vessels to constrict them. The increased volume and constriction help kick up your blood pressure. When that happens, the signals for the release of angiotensin end.

Finally, your kidneys are also responsible for producing the hormone erythropoietin, which talks to the bones and triggers production of red blood cells. Is there anything kidneys don't do?

When Kidneys Don't Work

If your kidneys fail, you don't have a way to excrete wastes. You don't have a mechanism for fluid and solute balance. You don't have a mechanism to keep your blood pressure under control. Any one of these things is ultimately and sometimes rapidly deadly. People with diabetes overwhelm their kidney function with sugars their body can't take up, which is why diabetes often is associated with complications involving blood pressure.

Circulatory and Respiratory

As you can see, there is a close connection among your digestive system, the wastes and extras that must be excreted from it, and your circulatory system, which is under constant control of the kidneys through fluid and solute balance. Throughout this chapter, the underlying—literally—component has been the circulatory system. Your small intestine cells absorb on one face and pass along nutrients to the blood at the other. Your kidneys associate closely with the blood to filter wastes and ensure fluid and electrolyte balance. So let's talk more about the circulatory system. This system is also closely tied to our respiratory system, all working to maintain homeostasis and engage in processes of material exchange.

Open or Closed?

You may be aware that your circulatory system is a closed network of tubes through which blood flows in one direction. If you were a mollusc (e.g., a snail or a slug) or an arthropod (e.g., an insect), your circulatory system would consist of a hemocoel, the blood-filled cavity described in Chapter 21. In these invertebrates, this cavity fills with fluid (bug blood) that diffuses back between cells and bathes the tissues. The fluid gets into the hemocoel thanks to that universal circulatory pump, the heart (or hearts), but the diffusion process is slow. The animal's movement helps keep the blood flowing.

Our closed circulatory system is different. Because of the blood pressure we maintain inside the vessels, the fluid distribution occurs much more quickly than it does by diffusion in an open system. The blood does not bathe body cavities; instead, as we have learned above, it travels past tissues in the vessels, allowing material exchange between the blood and the tissues. Vertebrates and a few invertebrates like the octopus and the earthworm have closed systems.

What does the circulatory system do, whether open or closed? Among other things, it transports gas and nutrients, takes wastes to be processed for excretion, totes hormones and signaling molecules to target tissues, and keeps body temperature level by moving heat around. Thus, it is a central player in both materials exchange and maintaining homeostasis.

As Circulation Goes, So Goes Respiration

You may have noticed that your blood is busy flowing by just about every tissue or system we've discussed. And that's true because it's a nice fluid delivery system for many things. One of the most important deliverables is oxygen gas, which you pick up by inhaling air into your lungs. Blood vessels adjacent to your *alveoli* grab the oxygen gas molecules thanks to diffusion—their blood is low on O_2, what you inhale is high in O_2—and tote it around to all of your needy tissues. And in you, *all* of your tissues need O_2. Hold your breath for a minute, and you'll understand that.

def•i•ni•tion

Alveoli (singular, alveolus) are clusters of air sacs at the ends of the maze of tubes that take air into your lungs. You have about 300 million alveoli per lung, and their moist surfaces facilitate gas exchange.

Now let's do a fun big-picture activity. Think back to those hazy days of Chapter 7 when we talked about cellular respiration. Do you remember which gas was a by-product of that process? If you recall

CO_2, you're right and you get a gold star. Well, that CO_2 has got to get out of you, or it'll kill you. Your blood obligingly drops off the O_2 and picks up the CO_2, running it by those lung tissues again. This time, the CO_2 diffuses from your blood into the lungs, and you exhale it. Phew. Got rid of that. If you're into circle-of-life kind of closure, you may now recall that plants will take that CO_2, use it in photosynthesis to make sugar, and give off O_2 for you to inhale.

Bio Basics

Respiration summary: inhaled $O_2 \rightarrow$ diffuses into blood passing lungs $\rightarrow$ goes to body tissues, diffuses into cells $\rightarrow CO_2$ diffuses out of cells $\rightarrow$ blood carries CO_2 to lungs $\rightarrow CO_2$ diffuses from blood into lungs $\rightarrow$ you exhale CO_2.

The Cardiac Cycle: Lub Dub, Lub Dub

Your blood doesn't just flow around of its own accord—it's got to have a pump. In us and other vertebrates, the pump is the powerful muscled organ called the heart.

We have a heart with four chambers. On top are two atria, the right atrium and the left atrium. On the bottom are two muscular ventricles, right and left. Let's follow some blood through these.

In our system, the rule is always that veins take blood *to* the heart, while arteries take blood *from* the heart. Some of our blood has traveled around the body and dropped off O_2, so it's coming to the heart deoxygenated. It enters the right atrium from the superior vena cava (a huge vein). Meanwhile, to the left, oxygenated blood fresh from the lungs comes to the heart from the pulmonary (lung) veins and enters the left atrium. The atria are now full of blood. It's time for a contraction, which is under involuntary nervous control. The period from the time of atrial filling to their contraction is called diastole.

Bio Basics

Blood in the veins (venous blood) is always deoxygenated *except* the freshly oxygenated blood traveling from the lungs through the pulmonary vein to the heart. Blood in the arteries (arterial blood) is always oxygenated *except* the deoxygenated blood that travels from the heart via the pulmonary artery to the lungs for more oxygen.

When the atria contract, the blood is pushed into the ventricles below them. The full ventricles then contract, a period called systole. The right ventricle contraction pushes the deoxygenated blood to the lungs for some more O_2. The powerful contraction of the left ventricle—possibly your body's most important muscle—pushes the oxygenated blood out to your body. Blood flow is kept moving in one direction thanks to one-way valves, flaps that open only in one direction and snap shut again when the pressure is relieved.

Blood Vessels, Flow, and Pressure

Your arteries, which carry the blood from the heart to the tissues, have muscular elastic walls that can adjust to increase or decrease pressure. One of the molecules that governs this adjustment is the angiotensin that also signals the kidney to alter fluid balance. Elastic arteries are important, and when heredity and lifestyle choices reduce this elasticity—harden the arteries—we suffer problems such as high blood pressure because the walls can no longer adjust as needed.

Bio Basics

Circulation summary: deoxygenated blood from body enters right atrium/oxygenated blood from the lungs enters left atrium→atria contract→blood pushed into right and left ventricles→ventricles contract→ right ventricular blood travels to lungs for O_2/left ventricular blood travels to body.

The muscular walls of the arteries help push the blood to tissues. Veins, on the other hand, which take blood away from tissues and to the heart, do not have these thickly muscular, elastic walls. Instead, they are thin walled, and we push blood upward through our veins against gravity thanks to contractions of our skeletal muscles. That's one reason it's a good idea to do point-and-flex exercises, for example, when you're on a long plane ride. It keeps blood flowing upward. Veins also have specialized one-way valves that keep blood from flowing backward.

Blood flowing away from the heart travels through vessels of decreasing diameter that we have classified as arteries, arterioles, and capillaries, which can be only one cell wide. The capillary beds or networks are where exchanges with tissues take place. Veins of smaller diameter are called venules.

The Least You Need to Know

♦ Your body uses many systems to exchange materials and maintain homeostasis, or internal balance.

♦ Your digestive system, from the mouth to the anus, achieves nutrient absorption and water balance.

♦ Your excretory system, from kidneys to urethra, maintains fluid and electrolyte balance.

♦ Your kidneys regulate blood pressure and produce erythropoietin, which stimulates red blood production from bones.

♦ Your circulatory system carries nutrients and gases throughout the body, using the heart as a muscular pump.

♦ Your respiratory system exchanges CO_2 for O_2, which diffuses to the circulatory system for delivery to cells.

Systems 2: Regulation, Reception, and Response

In This Chapter

- ◆ The nonspecific and specific defenses
- ◆ Hormone signaling from brain to body
- ◆ Nerve signaling
- ◆ The senses: see, feel, hear, taste, and smell

Your excretory, circulatory, respiratory, and digestive systems keep materials moving and maintain balance in your body. But no organism is an island. We all must encounter and manage an external environment while ensuring appropriate internal responses. We internally regulate our responses to messages we receive, and our internal interactions determine how we respond to those messages.

Immune System

Think about how many times in a day you touch things, so many different things that other people have touched. Money. Doorknobs. Countertops. Menus. Grossed out yet? Thankfully, we come equipped with defenses that manage to keep us pretty healthy in spite of the constant bombardment from other people's microbes. Sometimes, they fail, with illness or death as the result. But illness itself, such as fever, can be a sign that the defense is getting busy.

Nonspecific Defense: A General Response

Our first line of defense is our skin. Sure, it doesn't come across as that impressive—easily cut, even by paper, prone to drying out or getting too oily. But it's replete with antibacterial secretions (those oils!), has a low pH, and its thick layer of dead cells forms an excellent barrier against many common assaults (stop exfoliating!). Because it doesn't target anything specific, we consider it a nonspecific defense.

You know that green stuff that comes out of your nose? It's gross and disgusting, but it's another way your body successfully deploys a first-line defense against invaders. Saliva also knocks off microbes with tremendous success, and tears contain antibacterial compounds.

If your nonspecific external armory doesn't work and a pathogen slips past, you've always got your internal arsenal of nonspecific responses. Most of this arsenal consists of white blood cells. Macrophages, marauding white blood cells, zip around extending awesome cellular appendages to wrap around hapless bacterial invaders for ingestion. Natural killer cells approach cells tagged for destruction and shoot holes in them, triggering cell suicide.

When the attack is on, your brain receives messages from triggers called pyrogens (*pyro* = fire) that there's an assault and may respond by hiking up the heat. We call that a fever. While we don't like fever much, neither do many pathogens. Fever helps block bacterial growth, stimulates other aspects of the immune system, and boosts production of antiviral chemicals.

Specialized molecules called cytokines also take up arms, going to sites of invasion or to invaders and triggering inflammation with the intent to kill and heal. An injury that triggers cytokine release attracts specialized cells called mast cells that dump histamine at the damage site. Histamine makes capillary walls leaky, allowing other cells involved in defense to squeeze through and engage in the battle. Their activity can manifest acutely as swelling or redness. Over the long term, if inflammation persists, this defense itself becomes an illness.

The Immune Response: Specific and Efficient

All of these responses are nonspecific. They don't target a certain pathogen with a certain kind of weapon—they're just used against any pathogen, always in the same way. But what if these weapons fail? The body has more to try, the specific immune defense. This armory has two categories of weapon: cell-mediated defense and humoral or antibody-related defense.

Within this system, the body uses B cells, so-named because they mature in the bone, to produce antibodies (also called immuno-globulins), molecules designed to recognize a specific target. It uses T cells, which mature in the *thymus*, to stimulate the antibody-making cells and to kill off infected or diseased cells. Both cell types also have a memory version for quick recognition of a repeat invasion.

def•i•ni•tion

The **thymus** lies behind the breast-bone and is the site of T cell maturation. It usually winds down around puberty.

Cells of the Immune System: Naturally Killer

Keeping straight all of the different kinds of cells involved in defense can be, well, killer. First of all, they're all white blood cells, or leukocytes, primarily born in the bone marrow. These can be subdivided into granulocytes (neutrophils, basophils, and eosinophils) and agranulocytes (lympho-cytes, monocytes, and macrophages).

Bio Basics

Your skeletal system consists of bone and cartilage, a con-nective tissue. In addition to providing attachment for muscles and connectivity among tissues, the bones produce your blood cells from their source of stem cells in the living tissue.

Lymphocytes operate in specific immunity and include B cells, T cells, and natural killer cells. B cells occur as plasma cells that make antibodies and memory B cells that can quickly match antibodies to a specific repeat invader. T cells occur as cytotoxic T cells that kill infected or sick cells, helper T cells that alert other immune system cells, suppressor T cells that shut down the response when the battle is won, and memory T cells that remember the invaders, too. The table on the following page lists each cell type and gives its basic role in body defense.

Table of White Blood Cells

Cell	Features
B cells	Agranulocyte; lymphocyte; matured in bone (B!); plasma B cells secrete antibodies; memory B cells remember invaders and the correct antibody to make
Basophils	Granulocyte, release histamine
Eosinophils	Granulocyte, target parasites, bacteria; phagocytic; also primary cell involved in allergic reactions—will occur at high levels in presence of allergies
Macrophages	Agranulocyte, take up invaders, set up warning flags of invasion
Natural killer cells	Agranulocyte; lymphocyte; targets antibody-tagged cells, infected with viruses, from tumors; pokes holes with perforin; releases interferon (antiviral)
Neutrophils	Granulocyte; nonspecific defense; phagocytic; most common WBC in healthy people; 60-70% of WBCs; first on the scene
T cells	Agranulocyte; lymphocyte; mature in thymus (T!); cytotoxic T cells destroy "bad" cells; helper T cells alert B cells and cytotoxic T cells; suppressor T cells turn off response; memory T cells remember invader

A Closer Look: T Cells and B Cells

T cells are primarily involved in the cell-mediated response, so named because it achieves protection by killing infected body cells. The defender cells also target cancerous body cells. T cells get their name because they mature in the thymus, an organ that actually gradually disappears during adolescence.

B cells make antibodies. These molecules are proteins with the ends rearranged so that they have a specific molecular recognition of a specific invader. The rearrangement takes place in the DNA code for these proteins. Even a subtle change might mean a new kind of antibody that can recognize a new kind of invader. Once recognition happens, the antibody has a number of ways to let T cells know a cell is infected or to attract macrophages to an invading cell.

Antibodies: Structure and Function

Antibodies form five different groups. Each antibody consists of four protein chains, two long, or heavy chains, and two short, or light chains. The chains together assemble into the classic *Y* shape of the antibody. The rearrangement of amino acids takes place at the tips, with potentially millions of different possibilities. Because of their variability, these tip regions are called variable regions. The other regions are constant.

These molecules operate in defense in one of several ways. If they tag an invader through molecular recognition and stick to it, macrophages will identify them and kill the invader. An infected body cell may raise an SOS flag in the form of a molecule from the pathogen, called an *antigen*, and the antibody can tag these, targeting the body cell for destruction. They can also bind antigens on the invader, neutralizing its ability to attack body cells.

def•i•ni•tion

An **antigen** is anything that triggers antibody production— bacterial bits, viruses, toxins, etc.

How does the body keep from attacking itself? Healthy individuals fly "self" flags, groups of proteins called the major histocompatibility complex (MHC), that the immune cells recognize as "self." Disease can arise, however, if a mutation changes the MHC so that immune cells no longer recognize these protein flags as "self." The immune cells will attack these improperly flagged body cells, resulting in diseases categorized as autoimmune diseases (*auto* = self).

Endocrine System

This system is like your body's e-mail network, sending chemical messages along predetermined pathways to tissues capable of receiving them. Hormones are the messengers and can be either lipid or protein. Although people often think of the reproductive system when they think of hormones, many organs of the body participate in endocrine signaling and receiving, including the brain, thyroid, skin, kidneys, adrenals, gonads, gut, bones, and even the heart. In fact, I'm having a hard time thinking of a tissue that doesn't get involved in endocrine signaling in some way. Maybe your eyelashes? The table on the next page shows some of the major hormones, their sources, and their effects.

Some Hormones, the Tissues That Secrete Them, and Their Effects

Hormone	Secreted by	Effects
Aldosterone	Adrenal gland	Solute regulation, acts on kidneys
Estrogen	Ovaries	Sex development, uterine proliferation, ovulation
Follicle stimulating hormone	Anterior pituitary	Egg maturation/sperm production
Glucagon	Pancreas	Glycogen breakdown in the liver
Growth hormone	Anterior pituitary	Growth and development of body tissues
Insulin	Pancreas	Glucose uptake by body cells
Luteinizing hormone	Anterior pituitary	Ovulation or testerone secretion
Natriuretic peptide	Heart	Blood pressure control, blood solute concentration

Hey! Who Let These Hormones In?

Hormones usually show up because the levels of some molecule have dropped or risen above a threshold. These threshold-driven processes operate on feedback loops. Typically, the loop involves negative feedback: molecule A levels drop, a tissue responds by releasing hormone B, which triggers production of more molecule A, which reaches threshold levels and shuts down production of hormone B.

Most endocrine loops operate this way. The primary exception is childbirth. Oxytocin is the hormone that causes uterine contractions. The body's response to oxytocin's message actually triggers production of *more* oxytocin, leading to increasingly stronger uterine contractions. The uterus in pregnancy is the most powerful muscle in the body, and this positive feedback loop is necessary to make it contract enough to push a huge baby through a small, 10-cm (~4 inches) dilated, unforgiving cervix. Those of us who have experienced this process find it a bit uncomfortable, but that positive feedback loop sure does work well.

Brain Talk: No Endocrine System Without It

One of the organs that shows up frequently in endocrine signaling pathways is the brain, specifically two areas: the hypothalamus and the pituitary. Usually, the hypothalamus receives a message from some other tissue in the body that things are getting off target and passes the information along to the pituitary, which responds by sending out a hormone to get things back to normal. These pathways involving the hypothalamus, pituitary, and some third target tissue are axes. The third player in these axes might be the thyroid, the adrenal glands, or the gonads.

Brain Axes: Hypothalamus and Pituitary

The hypothalamus lies just above the brain stem and just below the thalamus (*hypo* = below). It connects the nervous system to the endocrine system and coordinates an astonishing array of inputs from both inside and outside the organism. In fact, most of the talking that the hypothalamus does with the pituitary involves sending along hormones that trigger the release of something.

The pituitary has two parts, the posterior and the anterior pituitary. In addition to releasing antidiuretic hormone, the posterior pituitary unleashes the power of oxytocin on the hapless uterus. But the anterior pituitary keeps busier. It sends out a laundry list of hormones targeting different tissues, including thyroid stimulating hormone, growth hormone, and follicle stimulating hormone, which stimulate, respectively, the thyroid, the body tissues, and the gonads.

When we get to reproduction, we wade into the mysteries of the hypothalamic-pituitary-gonadal axis. Here, let's meet the hypothalamic-pituitary-thyroid axis.

> ### Bio Bits
> Oxytocin also plays a role in breastfeeding and is known as the "trust" hormone because as a neurotransmitter it is associated with trust behaviors.

Your thyroid makes thyroid hormone, which regulates cell metabolism. When hormone levels fall, your hypothalamus registers the drop and responds with thyroid hormone releasing hormone sent to the anterior pituitary. The anterior pituitary responds by kicking out some thyroid stimulating hormone. This chemical messenger travels via the bloodstream to the thyroid, communicating with its cells and stimulating them to make more thyroid hormone. When the levels rise, the hypothalamus registers the increase and stops sending messages to the anterior pituitary—until hormone levels drop again. That's negative feedback.

Neural System: Touching a Nerve

Nervous systems can be as simple as a ladderlike network of neurons or as complex as the network of billions of brain cells that we've got concentrated in our skulls. Either way, the purpose is to receive information from inside or out, signal to the organism about its nature and intensity, integrate all related inputs (smell, sight, touch), and coordinate and initiate the response. That's all. No biggie.

We achieve this in spite of the fact that simply reading this book right now means you're getting inputs via your visual and touch pathways and sending outputs to your motor pathways, integrating sight, smell, sound (unless you're in a sensory-deprivation chamber), touch, and even taste from the millions of environmental and internal inputs that are bombarding you. Imagine what the world would be like if your nervous system couldn't sift through what's relevant and what isn't and help you sit there, calmly reading this book.

Don't Be Nervous: A Basic System Overview

The basic unit of the nervous system is the neuron, a cell specialized for conducting electrical impulses. Take a billion of these, and you've got the starting material for your brain, which consists of neurons and several other kinds of neural cells. The cerebrum of the brain has two halves, or hemispheres, that communicate through a central connection called the corpus callosum (tough body). The brain consists of an outer layer, the cortex, where all the important thinking processes take place. We've mapped the cortex and can pinpoint the areas of it where we integrate specific senses, such as the auditory cortex, where we register and integrate sound. The cortex is also called the grey matter because it consists of neuronal bodies, which are grey. Their axons, or extensions, reach inward from the cortex. Because of a fatty insulation on the axons, they look white, so we call this thick layer of tissue the white matter.

The brain communicates with the rest of our body by sending messages (except for those intended for the head) down the brain stem to the spinal cord, a bundle of nerves that talks with the rest of the body through nerve extensions from predictable levels all along the cord. The cord itself is surrounded by the bony vertebral column, which protects this highly sensitive tissue from insults.

Poke your arm. When you did that, your nerve endings at that spot sent a message through the sensory nerves, up through the spinal cord and to your brain. There, your brain registered, "My arm has been touched." If you had a super-hot index finger, your brain might then send a message back down the spinal cord through the motor neurons that activate your arm muscles, causing you to contract those muscles and move your arm away from the offending superheated finger. The figure gives an overview of how muscles contract, showing the contraction of a muscle unit called the sarcomere.

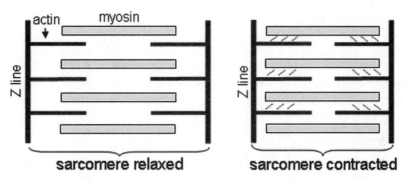

Nerve signaling triggers calcium release from the sarcoplasmic reticulum that triggers sarcomere contraction.

The brain and spinal cord make up the central nervous system, which integrates sensory input and motor output. The rest of your nervous system, the part that lets you sense and move, is your peripheral nervous system. If you wiggle your big toe right now, your central nervous system sent a message through the motor neurons of the spine to the peripheral nervous system saying, "Wiggle your toe." Your brain is in charge of memory, learning, emotion, integration, and response, while your spinal cord passes along messages like a two-way highway and is the seat of your tendon reflexes.

Your peripheral nervous system also is a two-way highway, consisting of a set of sensory nerve extensions for taking information in and extensions of motor nerves from the spinal cord for sending instructions out. The motor nerve extensions themselves fall into two categories. Those that talk to your skeletal muscles and control voluntary movement are your somatic nervous system. Those that oversee your involuntary muscle activity are your autonomic nervous system.

This latter bunch divides further into the sympathetic and parasympathetic autonomic nervous systems. The sympathetic nervous system shuts down all but the mission-critical muscle activity for a fight-or-flight response. That's great if a bear is chasing you but gets in the way if you have this response when faced with public speaking. Your parasympathetic nervous system oversees your "rest and ruminant" activities—the resting stuff that goes on while you're lying around, presumably chewing cud or something.

Cell Support: Neurons and Neuronal Helpers

While the neurons do all the communicating, they receive support from other cells that nourish and protect them. But first, let's talk about the neuron.

A neuron has three basic parts. The cell body, or soma, is home to the nucleus and organelles. Sticking out all over the soma are pointy extensions called dendrites (*dendr* = trees). Extending from the soma is the single elongated axon that ends in an axon terminal, where the neuron interacts with another nerve cell's dendrites or communicates with a tissue, like muscle.

Neurons fall into three basic categories, not including some specialized types that we won't get into here. Afferent neurons carry information from the body to the central nervous system, so they are sensory. Efferent neurons carry information the other way, so they are motor. Finally, the interneurons act as neuronal connectors within the central nervous system, linking sensory and motor neurons.

Neurons don't do it all alone. They've got a supporting cast of cells, including glial cells and myelin. Glial cells (also known as glia) are so important that there's at least one per neuron throughout the brain, where they occur. They protect neurons and provide them nutrition and were even once known as the glue of the nervous system (*glia* = glue) because they hold neurons in place and buffer them from other neurons. Wouldn't want those nerves sliding around all over the place, now would we?

Glia can occur as astrocytes, which take care of most of the jobs described above, and as microglia, which dispose of dead neurons. Oligodendroglia produce the myelin that insulates the neuronal axons in the central nervous system. Without the fatty myelin insulation, electrical impulses through the nerve would bounce all over the place, rather than passing rapidly and smoothly down the axon.

Schwann cells do the same job in the peripheral nervous system that the oligodendroglia perform centrally: they produce the axonal insulation called myelin that ensures rapid transmission of the electrical signal.

Touching a Nerve: What Happens When You Do

Poke your arm again. We're about to get into what happens when you trigger those axon terminals just under your skin. It all begins with an imbalance.

A Resting Membrane Is an Imbalanced Membrane

Neurons must be constantly at the ready to receive and perpetuate electrical signals. They maintain readiness by keeping the area just inside the cell membrane in a state of electrical and chemical imbalance relative to the outside of the membrane. The cell stays negative on the inside by pumping out positive ions at a greater rate than it lets them in.

The ions involved in maintaining this negative potential of the inner membrane are sodium (Na+) and potassium (K+). The cell has two things to consider in terms of its ions: their chemical balance and their electrical balance. Ions have a charge, and as nature abhors a gradient, the effort is toward charge equilibrium on either side of the membrane. But they also have chemical properties, and the cell strives to achieve a chemical balance, too. The upshot of trying to balance the chemical and the electrical components inside and out is an imbalance.

There is more K+ inside than outside. Simple passive movement allows K+ to leak out of the cell, seeking equilibrium. However, as these positively charged ions take their + with them, they leave behind a + deficit, making the inside of the cell membrane more negative. The cell, to correct this negativity, goes to the trouble to pump K+ *back in*. Yes, as K+ leaks out down its concentration gradient through a few open channels, the cell pumps it back in, apparently confused over the dual demands of chemical and electrical balance.

Meanwhile, something strange is going on with Na+. There's lots outside the cell and relatively little inside. Naturally, tons of Na+ is knocking at the cell membrane doorways, clamoring to get in. The cell, rather than opening the gates to Na+, actually pumps even more Na+ out of the cell. In fact, the same pump that brings K+ in pumps Na+ out. This sodium-potassium pump boots out three Na+ for every two K+ it pumps in. Yes, that's three + out for every two + in. The result? A net negative on the inside. The fact that the cell also has a lot of large, negatively charged proteins lying around inside just adds to the atmosphere of negativity in there.

def•i•ni•tion

The **membrane potential** is the voltage difference between the inside and outside of the membrane.

What does that leave us? A cell with K+ clamoring to get out and Na+ banging on the gated channels to get in. And there's a voltage differential across the membrane, with the inside resting at a nice, -70 mV (millivolts) compared to the outside. Believe it or not, this resting *membrane potential* is the perfect setting for some nerve-signaling action. Action potential, that is.

Action Potentials

Poking your arm triggers an electrical message that passes along the sensory circuits to your brain. What you've really done is kick the neurons out of their comfortable -70 mV resting potential. You've triggered the opening of some of those Na+ channels. As the Na+ moves into the cell, the resting potential becomes more positive, a phenomenon called depolarization.

This trickle of positive Na+ sunshine into the cell may continue until the inner membrane depolarizes to a certain standard threshold, about -55 mV. When that happens, the cell has hit the all-or-nothing point and fires off a true action potential. The Na+ floodgates fly open, one after the other like falling dominoes, and Na+ rushes freely and rapidly into the cell. This sudden influx of + rapidly alters the cell's inner potential to about +30 mV. The Na+ channels will continue to open in a wave all along the membrane and down the axon, so that the depolarization continues all the way to the axon terminal.

Transmitting the Electrical Message

What happens next? Well, the electrical signal doesn't just leap from one neuron to the next (an exception is the heart). First of all, there's a space between the axon terminal and the next neuron called a synaptic cleft. The neuron on the sending side is the presynaptic cell, and the neuron on the receiving side is the postsynaptic cell. In the axon terminal of the presynaptic cell lie vesicles packed with signaling molecules called neurotransmitters. Examples of neurotransmitters include serotonin and dopamine.

When the wave of the action potential reaches the axon terminal, it triggers the vesicles to release their neurotransmitters. These signaling molecules diffuse rapidly—we're talking millisecond rapidly—across the synaptic cleft to the waiting postsynaptic neuron. There, receptors recognize the neurotransmitter's signal, which triggers opening of more Na+ channels, allowing the Na+ to flood into this next cell, depolarizing it and continuing the electrical signal. Action potentials in a new cell begin in the dendrites.

So how does it all end? How does the body repolarize the cell back to the negative resting potential? Remember the K+, sitting there in the cell, dying to get out but always being pumped back in? Now is its turn to shine. The Na+ gates snap shut as quickly as they open. With the passing of the depolarization wave, K+ channels in the cell membrane now fly open. The waiting legions of K+ finally get to rush out, taking their + charge with them and leaving the inside of the membrane back at its negative resting potential. These K+ channels close, and the sodium-potassium pump once again begins its robotic exchange of two K+ in, three Na+ out. Meanwhile, K+ continues its stealthy, low-level leakage across the membrane, wherever it can find an opening. The cell is back at rest at -70 mV.

> **Biohazard!**
>
> Although I keep talking about electrical signaling, don't lose track of the fact that chemical imbalances, such as specific types of ions in nonequilibrium inside and outside the cell, are key to nerve signaling.

Sensory: Did You Feel That?

The messages that travel the sensory pathways arrive via the sensory nerves. But how does, say, the smell of bacon get to your brain?

Mechanical and Chemical Sensing

You have a couple of ways to trigger an action potential. Poking your skin is a mechanical approach. Smelling delicious food sends chemical messages to the nerves that set off an electrical signal by triggering opening of the Na+ channels.

Where Am I? Sensing Body Position

Proprioception (*proprio* = self) is our sense of where we are in space. We have specific receptors that let us know what we're around and what's around us. Without these receptors triggering electrical messages to the brain, we'd constantly be bumping into everything. We also sense our position against gravity thanks to our vestibular system, which relies on signals from the inner ear to tell us our orientation.

Taste and Smell: Inseparable Senses

Chemical signaling is the basis of taste and smell. Smells are particulate, meaning that actual, physical molecules bind to receptors on chemosensory cells at the top of our nasal passages. Think about that the next time you're in a chemical toilet. (Eww!) The binding triggers a signal along the nerve pathways to the brain, which integrates the information and initiates a response. Tastes—which fall into five basic categories—operate much the same way; in fact, much of what we taste we actually process through smell. The five tastes are sour (acids), sweet (sugars), bitter (alkaloids, designed to warn you of toxins), salty (mmmm), and umami, which refers to savory tastes in foods like meats and cheeses (mmmm, again).

Hearing: What Hair Has to Do with It

Your ear is constructed to take a sound wave and transmogrify it into an electrical signal. It funnels sound waves from the outer ear to a membrane called the eardrum. The vibrating membrane transfers the vibrations to the bones of the middle ear, the hammer, anvil, and stirrup. The vibrations then transfer to the inner ear, which is full of fluid. There, a tube called the cochlea lies coiled up like a French horn. It picks up the vibrations in its organ of Corti, which has cells lined with little structures we refer to as hairs. When the vibrations move the hairs, an action potential is set off. We translate the message we receive in the brain as "sound."

Seeing: Do You See What I See?

We see because we register and integrate the presence of light. Our eyes are under protection from a tough outer layer called the cornea. Light first must pass the cornea, then onward through that dark center we call the pupil. As you may know, the pupil can change in size depending on how light or dark the environment is—less light = larger pupil. The color around the pupil that we all think of as so significant to us as individuals is called the iris, and it controls the pupil's size.

Bio Bits

When we blink, we gradually wear away our corneas. The cornea replaces the tissue using a pool of stem cells it has tucked away.

After light passes through the pupil, it focuses through the lens of the eye onto the retina at the back of the eyeball. In the retina are the cells that register the light and translate it into an electrical signal that zaps the brain with visual information. The cells are the rods, which help us see in dim light, and the cones, which are in charge of our color discrimination. Their signals aggregate and travel via a good-size pathway called the optic nerve to the visual cortex, where the brain integrates them, which we perceive as an image.

The Least You Need to Know

◆ Your nonspecific defenses include skin, secretions, fever, and marauding cells like macrophages and natural killers.

◆ Your specific defenses involve B cells, which make antibodies, and T cells, which manage infected cells.

◆ The endocrine system operates on feedback loops, including axes involving the hypothalamus and pituitary.

◆ Your nervous system consists of the central brain and spinal cord and the peripheral sensory and motor nerves.

◆ Nerve cells use electrical and chemical imbalance to facilitate electrical communication via action potentials.

◆ We sense via stimuli (chemicals, touch) of action potentials in sensory nerves that send messages to the brain.

Chapter 24

Animal Development

In This Chapter

- ◆ Sexual and asexual reproduction
- ◆ Female and male gamete formation
- ◆ Fertilization, implantation, and early development
- ◆ Major processes in early development
- ◆ Human development by trimester

Living things reproduce. It ensures continuation of the species. But while billions and billions of the planet's organisms reproduce asexually, either by binary fission or mitosis (see Chapter 10), some, including us, use sexual reproduction. The result is an offspring with a unique genome. A sperm's fertilization of the egg produces the first cell of this nascent organism. What happens from there is the process of embryogenesis and development, in which that single cell ultimately produces from a thousand to billions of cells, depending on the organism—cells that have differentiated based on developmental programs that have been under refinement for millennia.

Sex and Reproduction

Scientists argue over why sexual reproduction exists. Is it beneficial because it shakes up the genetic code, giving nature a greater number of choices? If that were the case, why is it that the most long-term taxa on the planet, prokaryotes, rely on asexual reproduction, obviously with great success? No one has quite nailed down the answers.

Bio Basics

Asexual reproduction involves one parent, no gamete fusion, and offspring that are usually identical to the parent. Sexual reproduction involves two parents, gametes fusing, offspring that are a mix of the two parents, and gene shuffling.

Sexual or Asexual? Some Organisms Can Choose

While we use sexual reproduction to make new individuals, some organisms can select one or the other, based on circumstances. For organisms that have a choice, often the determinant is the environment. When the going gets tough, the asexual may become sexual. Again, this could be a way of introducing some variation into a difficult and highly choosy environment, but that explanation, while persuasive, remains hypothetical.

Bio Bits

A familiar organism that uses asexual and sexual reproduction is the ant. The queen's unfertilized eggs will develop into haploid males, while the fertilized, diploid eggs will develop as females.

Roughly categorized, all vertebrates, with a few notable exceptions, reproduce sexually. Insects and other invertebrates are largely sexual, although social insects like ants may use both sexual and asexual reproduction. For many sexually reproducing species, the key players are some form of male gamete, sperm in animals or the nuclei of pollen in plants, and an egg designed to recognize and fuse with that sperm.

Sperm Formation and Delivery: The Male Reproductive System

Our discussion of sex in this chapter focuses primarily on people. One key feature to note about humans and many other mammals is how different male versus female meiosis is.

Bio Basics _____

Male meiosis takes us back to the hypothalamic-pituitary-gonadal axis (Chapter 23). Low testosterone output from gonads triggers hypothalamic release of hypothalamus gonadotropin–releasing hormone, which signals the anterior pituitary to release luteinizing hormone and follicle stimulating hormone, which stimulate Leydig cells to make sperm.

A male produces sperm in long, stringy structures called seminiferous tubules, located in the testes, which themselves lie hanging in a pouch called the scrotum. Leading from the testes to the urethra, where sperm exit, is the vas deferens, a sperm delivery tube. The prostate gland, Cowper's gland, and the seminal vesicles produce the nutrition that the sperm need to thrive.

If you cut a slice of seminiferous tubule (ouch), you'd see an assortment of cells. Spermatogonia, destined to undergo meiosis and produce sperm, lie in a circle around the tube. Interspersed among them are the Sertoli cells, which support the spermatogonia, providing nourishment. And along the outside of this cellular circle lie the interstitial cells (also known as Leydig cells), which produce the testosterone (in response to stimulation from an anterior pituitary hormone called luteinizing hormone) that triggers meiosis.

Males don't start making gametes until hormone signaling in puberty kicks off the process. But when they get going on gamete production or spermatogenesis, they really get going, producing four viable sperm with each meiotic event and making millions and millions of little swimmers at a time. Of course, the male gamete itself is usually quite different from the female version and a lot smaller, a state called heterogamy. In humans, the egg is about 20 times larger than the sperm. It's also immobile, whereas sperm can swim (and pollen can float or get stuck on a bee's behind).

Bio Basics _____

Development of sex organs starts from the same structures in males and females. A protrusion called the genital tubercle usually develops as either a penis or clitoris depending on hormone signaling. Likewise, the developing embryo usually forms a scrotum or labia from shared starting tissues depending on the hormonal environment.

Egg Formation: The Female Reproductive System

Females are more complex (as any self-help book would try to tell you). As we learned many chapters ago (Chapter 11), the result of female meiosis, called oogenesis, is usually one big fat, viable egg and three teensy polar bodies.

Egg development takes place in the ovary, the female gonad. The ovary consists of two basic parts, the cortex and the medulla. The egg traces its path along the cortex, or outer part, while the medulla is home to the blood vessels that bring oxygen and nutrients to the tissues.

As the egg develops its way through the cortex, it does so inside a follicle that develops along with it in this follicular phase. About halfway around the ovary, the egg will pop out into the fallopian tube, which leads to the uterus, where any resulting embryo will implant. This release, called ovulation, leaves the follicle behind, which then becomes the corpus luteum (yellow body).

> **Bio Basics**
>
> The birth control pill usually provides a steady, low dose of estrogen and progesterone, which signals to the brain and suppresses hormone fluctuations, thus preventing ovulation.

Girls are thought to be born with their eggs already in the process of formation. In fact, those eggs started forming when the girl was an embryo, beginning meiosis but getting stuck in prophase of meiosis I in the ovary. These primary oocytes go no further in meiosis until puberty, when hormones kick in and stimulate one primary oocyte to cycle through the ovary. One of those hormones is estrogen, and it also stimulates the uterine lining to thicken in anticipation of an arriving nascent embryo. As the oocyte cycles through the ovary, it completes meiosis I and begins meiosis II.

> **Bio Bits**
>
> If implantation into the uterine lining occurs, the body will continue making progesterone to support the lining. Then the embryo takes over and starts making human chorionic gonadotropin, which supports the uterine lining. This is the hormone that pregnancy tests from the drugstore detect.

About midway through a typical cycle, the ovary fires the egg into the fallopian tube, driven by a huge spike in estrogen (this is ovulation). The egg lingers there for about a day or less, awaiting a swimming sperm, and stuck in metaphase of meiosis II. After the egg exploded into the fallopian tube, the corpus luteum pumps out progesterone to keep the uterine lining in place in case a fertilized egg shows up. If nothing arrives in the couple of weeks after ovulation (a phase called the luteal phase), progesterone and estrogen levels drop, and this lining sheds. We call this event menstruation, which marks the beginning of a new cycle.

Fertilization and Implantation

The window for fertilization in humans is about a day. If fertilization does occur, the reaction is immediate. The egg swiftly completes meiosis, kicking out another polar body, while the nuclei of the sperm and egg start migrating toward one another to fuse. Cell division begins quickly so that by the time the whole package arrives at the uterus, it's already in the blastocyst stage, consisting of about 100 cells.

So the egg after ovulation sits there, stuck in meiosis II, waiting for an army of sperm to show up. But if the little swimmers do arrive, nature has decreed that only one will succeed in getting in.

> **Bio Bits**
>
> Fertilization timing can vary: In us, it's in the middle of meiosis II in the egg. In sponges, some round-worms, and, strangely, dogs and foxes, it's at the primary oocyte stage. In sea anemones and sea urchins, it's after egg meiosis is complete.

The Basics: One Sperm Only, Please

It's rare in animals for more than one sperm to fuse with an egg, and it's also usually fatal. Nature has gone to a lot of trouble to ensure that once a single sperm has fused with the egg, nobody else can get in.

The first step in fertilization is simple recognition. The structure responsible for species protein-protein recognition between two gametes is called the zona pellucida (clear zone) in mammals, which surrounds the egg and consists primarily of glyco-proteins. If you recall, these are the kinds of proteins most often involved in cell-cell recognition.

In mammals, the female reproductive tract and the sperm interact to bring the sperm to final maturation. This swimming male gamete consists primarily of its nucleus packed into a cell membrane with scant cytoplasm; a head called the acrosome that's filled with destructive enzymes; and that spinning, swimming flagellum, driven by a mitochondrion.

Upon reaching the egg, the mammalian sperm binds to the outside, triggering imme-diate release of acrosomal enzymes, called the acrosomal reaction. These enzymes chew a hole into the egg, at which point the membranes of the two gametes fuse. Now the two nuclei begin migrating toward one another in the cytoplasm.

If recognition is successful, then the first arriving sperm at the egg, in the midst of its tremendous success, is also in for a big surprise, a sort of egg explosion going off all around it.

The Processes Compared: Urchins, Frogs, and Mammals

People who study the processes of an organism's development from fertilization onward are called developmental biologists. Developmental biologists usually work with a standard set of species, or animal models, that for various reasons became the go-to animals for such studies. Among these is the sea urchin, so now, we're going to talk about the sea urchin.

Sea urchins and frogs prevent entry of more than one sperm in a process called the fast block to polyspermy. To understand this, you're gonna have to recall membrane depolarization (Chapter 23). When the winning sperm binds the outside of a sea urchin or frog egg, the egg immediately responds by opening the floodgates to Na+. The Na+ rushing into the egg depolarizes the inside membrane—did you see that coming?—producing a total electrochemical rejection of any other sperm. As with any depolarization, this block is temporary. The membrane will repolarize (K+ leak, anyone?). Theoretically, at this point, more sperm might be able to fuse, but presumably the egg will have moved on to other things.

Sea urchins also use the slow block to polyspermy as a backup system, while mammals use only the slow block. This block relies on a massive release of destructive proteins. When the sperm fuses, the egg responds with an explosion of calcium from special granules. The calcium facilitates release of destructo-proteins called proteases, which wander around outside the egg and chew up all of those sperm-egg recognition glycoproteins. No glycoproteins, no more recognition, no more sperm entry.

The zona pellucida (also known in other animals as the vitelline membrane) then separates from the plasma membrane, and water flows into the gap. The now-swollen vitelline membrane has become a fertilization envelope. Without recognition proteins and with a puffy, fluid-filled barrier in the way, no other sperm will access the egg. While you can see how the electrochemical or fast response is a lot faster, the slow response is certainly more permanent.

Gastrulation: Forming Tissue Layers

We got into gastrulation in Chapter 21, the process of forming three initial tissue layers from which all other tissues arise. Those tissue types are the endoderm, mesoderm, and ectoderm. We know the outcome of gastrulation, but here, we're going to spend a little time talking about how we get to that point. The table shows some of the stages, players, and processes of earliest embryonic development and the confusing names we have for them.

The Terminology of Embryonic Development

Term	What It Means
Blastomere	A cell that results from cleavage of the zygote.
Blastocoel	The cavity that forms in the blastula and allows gastrulation to take place by making space for cells to move around.
Blastocyst	The mammalian version of the blastula. Why? Because mammals are just that special.
Blastopore	The invagination in the blastula where gastrulation begins. These cells invaginate into the blastocoel. In mammals and other deuterostomes, the blastopore is destined to become the anus.
Blastula	A ball of blastomeres; the developmental stage at implantation, about 100 cells.
Gastrula	The embryo after gastrulation.
Morula	The embryo at the "mulberry" stage (*morula* = mulberry), consisting of about 16 blastomeres. Just prior to the blastula/blastocyst stage.
Neurula	The embryo after neurulation.

Once the sperm and egg fuse, the resulting cell is called a zygote. The first thing a zygote does is begin the process of cleavage, cell division and cytoplasmic division that results in two and then four and then eight identical cells.

Except that they are not entirely identical. Recall that all of the cytoplasm of this newly formed zygote came from the egg. The egg has come prepared and has a cytoplasm packed with molecules at the ready, called cytoplasmic determinants. These mysterious operators, usually mRNAs ready for translation, are charged with assorting into the newly forming cells and directing subsequent processes until the embryo itself is ready to take over. Because these cytoplasmic determinants don't assort identically, those cells from the first division are not truly identical. Think of taking a bunch of raisins and sprinkling them randomly on top of a pie. Now cut the pie into four equal pieces, which reflects how those first few divisions divvy up the cytoplasm. Sure, the pieces are the same size, but is the raisin distribution the same? Likely, it is not.

Eventually, the embryo is going to have to cut the apron strings and do everything on its own. Depending on the organism, the embryo will spend a certain amount of time relying on the egg and those determinants to direct early development. But at some point, a shift takes place. Because the stage at which this shift typically occurs is the blastula stage, we call it the mid-blastula transition (MBT). At this point, the embryo's own genetic code takes over.

The MBT is a major turning point for the developing organism. If there's a catastrophic problem with its own genetic code, then development may fail completely. The biggest test of the ability of the embryo's own genes is gastrulation, which quickly follows this embryonic takeover. For this reason, developmental biologists consider gastrulation to be the most important moment of an organism's life, a true test of the viability of this newly created genome.

Neurulation: Beginning a Nervous System

The cells that will become your central nervous system begin life on the outside of the developing embryo. Very early in development, right around the time of gastrulation, the process of CNS formation, called neurulation (*NYUR-u-lation*), begins. This is also the process in which the cells, called neural crest cells, arise. These cells migrate to many different areas of the body and become neurons and some other tissues of ectodermal origin.

Neurulation begins on what we might call the "top" of the embryo, along what eventually will become its back (technically its dorsal area). The first step is a thickening of the ectoderm in this area. This thickening is called the neural plate. Think of putting a plate made of soft dough on the back of a ball. The next step is that the sides of the dough plate start to curl upward and meet over the center of the plate to form a tube. In this way, the cells that were on the outside now end up inside, and the cells that end up on the outside become epidermis, covering and protecting the central nervous system. This tube is called the neural tube.

Biohazard!

If you're female, watch your folic acid intake, even if you have no intention of becoming pregnant. Foods are now fortified with folic acid because of its role in complete closure of the neural tube. Defects in this closure are related to developmental anomalies including anancephaly (no brain development) and spina bifida. Studies indicate that adequate folic acid intake even before conception is important.

Where the edges of the folds meet, these cells will become neural crest cells (because they're at the crest of the plate edges). The tube itself will become the brain and spinal cord. At the "head" end, the brain will form, while the rest will become cord. All along the neural tube, paired structures called somites form. These segments reflect the segmentation of the vertebrate and will predictably give rise to the different segments of the torso, the vertebrae and ribs, muscle, and skin.

Further Development: The Developmental Map

How do we know all of this, this business about which cells will eventually give rise to different tissues? Developmental biologists have carefully monitored specific cells using various techniques to follow their fates. The overall map of the fates of different cell lines in an organism is called a fate map.

Nematodes: The Completely Mapped Worm

Most organisms are so complex and the fates of their germ cell lines are so various that mapping the ultimate origin of every cell in the organism is well nigh impossible. At least for now. But the nematode *Caenabdoritis elegans* presented a golden opportunity because this tiny roundworm consists of fewer than 1,000 cells. It also happens to be—along with the sea urchin, zebrafish, African clawed frog (*Xenopus laevis*), and the chicken—one of the standard animal models of embryology and developmental biology.

Thus it came about that scientists have indeed mapped the fate of every single one of the cells in *C. elegans*. Why are such maps significant? From the very first cell, we can analyze its components, position, interactions, and behavior to determine what factors contribute to its being fated to become what it becomes in the organism.

> **Bio Basics**
>
> Because of space limitations, we're focused on direct development, the growth of an adult organism from a juvenile organism that looks a lot like it. But many organisms instead exhibit indirect development in which the juvenile and adult stages are very different. The most recognizable examples of this kind of development are caterpillar to butterfly and tadpole to frog.

Us: Not so Complete

As handy as a fate map is, for most other organisms, they must remain, perforce, impartial. We have partially mapped the fate of the cells or a group of cells in many organisms, but complete fate maps are a rarity. Human development still retains many mysteries to unlock. Because mammalian development is "hidden," i.e., does not occur where we can easily visualize it the way a frog or a sea urchin does, it's more difficult for us to tease apart and experiment with. Because of the obvious ethical issues surrounding any use of human embryos in an ongoing developmental process, we rely on mouse models for such studies. Even these, however, prove difficult for visualization, especially the kind of real-time observation we can get from the more readily available invertebrate and amphibian embryos.

Human Development: By Trimester

From fertilization, complete human development takes about 36 to 40 weeks. For the first seven days or so, the dividing ball of cells makes its way through the fallopian tube to the uterus, where it ultimately implants in the uterine wall about day 7 to 10, at the blastocyst stage. During the first couple of months of gestation, we refer to the developing organism as an *embryo;* after that, it is a *fetus.*

def•i•ni•tion

In biology, an **embryo** is the developing human from conception to the end of the eighth week of development. A **fetus** is the developing human from the ninth week until birth.

In the first third, or trimester, of human development (weeks 1–13), gastrulation and neurulation complete early on. A specific group of cells in the embryo is destined to become the placenta, which participates in nutrient and waste exchange with the maternal blood supply via the umbilical cord. Another group of cells will make up the sac, or amnion, that holds the fluid to cushion the developing embryo and fetus throughout gestation. A tiny flash of a heartbeat is visible by ultrasound as early as six weeks; by this time, the tube of cells that will become the heart has formed and started the electrical signaling that causes muscle contractions. By the end of this trimester, the major components of the organ systems will have differentiated. There will be distinguishable limbs, fingers, and toes, and the fetus will be about three inches long. By this time, the head and body proportions are similar to what we see in newborns.

The second trimester (weeks 14–25) is dedicated to maturation of some systems, including outward structures and processes such as movement and reflexes. It is usually sometime in this trimester that the mother feels the first fetal movements. Eye development completes, and taste buds appear. The fetus can suck a thumb, swallow, and hear your voice. About the 23rd week of development is the time that the fetus has some probability of surviving, if born, although the odds remain extremely low. Even with survival, potential complications from such a premature emergence are highly likely. By the end of this trimester, the fetus will weigh in at about a pound and be about a foot long.

The third trimester is all about growing and laying down fat, which is not only good for the body but also for the fetal brain. Lung development, which remains immature until well into this trimester, completes, as well. At birth, the average baby in the United States weighs about 7.5 pounds and is about 20 inches long. These values vary widely, however. I know of a perfectly healthy full-term baby who was under 6 pounds and another full-term baby who emerged weighing a jaw-dropping (and other painful things) 10 pounds, 11 ounces.

The Least You Need to Know

◆ Sperm arise from spermatogonia in the testes via spermatogenesis, with support of interstitial and Sertoli cells.

◆ Females go through a follicular phase, ovulation, and the luteal phase, when the corpus luteum produces progesterone.

◆ Male and female gamete production are regulated by hormone feedback via the hypothalamic-pituitary-gonadal axis.

◆ Organisms prevent entry of more than one sperm into the egg by fast (electrical) or slow (chemical) blocks or both.

◆ Cytoplasmic determinants oversee early development until the mid-blastula transition, when the embryo takes over.

◆ Neurulation results in the early embryonic structures that eventually develop into the central nervous system.

Part 6

Ecology

Behavior is a response to a stimulus, and no organism lives without stimuli. In fact, almost all species live and interact with other species, forming communities that live, mate, play, and eat (sometime each other) together. The populations of each species come together to form the communities that interact with each other and the nonliving environment to form the ecosystem. The functions of life facilitate cycling of nutrients through these systems, while the interaction of nonliving forces like precipitation and temperature shape who gets to live where and with whom. It all starts from bare rock, but it ends with the network of interconnected ecosystems we call the biosphere.

Chapter 25

Animal Behavior

In This Chapter

- Genetics and behavior
- Learning and behavior
- How and why animals communicate
- The truth about altruism

Why are some people cranky while others seem preternaturally sunny? What is the deal with terriers and their feistiness? Why does the smell of baking brownies make us drool? Behavior is a complex phenotype because of its complex origins in genes and environment and the network of neural and hormonal signals that come together to produce it.

What Is Behavior, and How Do We Study It?

People study behavior from a few different perspectives. Scientists addressing it from a biological viewpoint are called ethologists. Those who delve into the learning aspects of behavior are experimental psychologists, while those who address the evolutionary backdrop of behavior are behavioral ecologists.

Regardless of their scientific pigeonhole, they all study how organisms react and act in response to their environment. These reactions fall under the control of the nervous system, which transmits those messages inward via the sensory pathways and then sends out motor pathway messages for response. But as you may be aware, not every incoming message is created equal. Sometimes, even the exact same input can register differently. What upsets you terribly one day might be something you can brush off the next. The system that modulates these variations in perception is your endocrine system. Thus, behavior has two masters, the nervous system and the endocrine system, that engage in a great deal of crosstalk with each other.

Given the involvement of these systems, it may come as no surprise that we study the underlying causes of behavior by focusing on endocrine and neural pathways. We study these in different contexts, including the natural environment and experimentally induced situations in which we've altered either the external environment or internally tweaked the neural or endocrine pathways associated with the target behavior. For example, if we're interested in the influence of testosterone on sex-specific behavior, we might expose female animals to testosterone through injection, altering their internal environment to see if this endocrine interference changes their behavior. By the way, it does.

Genes and Behavior: Nature to Nurture

How much of behavior lies in the genes (nature) and how much lies in the environment (nurture)? I've always viewed the nature versus nurture argument as a bit of a false dichotomy because no gene is an island: all genes are likely subject to multiple environmental influences, so even if you have the exact same genetic sequence as someone else, environmental influences can alter how those genes are expressed, or what your body does with the proteins after they're made. The possibilities for fine tuning of what lies encoded in the original sequence are legion.

Behaviors can be classified into phenotypes, just like hair color or height. The inference to draw from this information is that behavior ultimately lies in the genes, the genotype that underlies the phenotype. Thus, the raw constructs of your personality—of everything you do, the way you respond to the world around you—lie sketched out in your genome. But the path from gene sequence to expression to protein behavior carries many forks, loops, and alternate byways so that even with identical nature (genes), nurture (environmental influences) can still produce subtle or even notable differences.

Identical twins illustrate the influence of environment on genes. They are identical in their gene sequences, each having arisen from one of the two cells that emerged from the first division of a zygote. If you've ever met a pair of identical twins, you know that they do look quite similar. They may also behave quite similarly. Fascinating studies of identical twins separated at birth have revealed strong similarities in the most unexpected ways, including, for one pair of identical twins who after birth didn't even meet until middle age, a tendency to read magazines back to front. Is that genetic?

But studies of identical twins also clearly demonstrate that they start with some nongenetic differences and become even more different with age. One twin may develop a health disorder that has a strong heritable (genetic) component, yet the other twin does not. Evidence suggests that identical twins accumulate epigenetic (above the gene) changes that differ. Usually, these changes involve addition of chemical tags on the gene, which may silence it or turn it on. These epigenetic changes can be akin to having a different genome—if the chemical tags suppress gene expression, for example, then it's pretty much like not having that gene at all. In this way, even people with identical gene sequences become different as their patterns of chemical tags become more different with age.

So nature versus nurture? Really, it's a mix of both, but it all starts with nature.

> **Bio Bits**
>
> The nongenetic differences that accumulate over time involve the addition of methyl (CH_3) groups to parts of the genome. Because these changes can alter how or if a gene is expressed, they can essentially have the same effect as changing the gene itself.

Learning and Behavior

Some behaviors are innate, hardwired into the genes and happen in the absence of any kind of learning. Fawns start walking within minutes of birth. Dogs bark. While walking may undergo modulation thanks to environment (don't step on that thornbush!) and the bark can change based on the situation (squirrel!), the behavior is something that just … happens. No tutorials required. The genes have it.

Some innate behaviors are highly stereotyped, which means that if the stimulus or trigger is present, the behavior will manifest—no matter what. Thus, if you scratch your dog in just the right place on his shoulder blade, he'll do that phantom scratching thing every time. We call these stereotyped behaviors fixed action patterns because they're fixed actions in the same pattern always in response to the same stimulus.

Innate behaviors often carry a clear evolutionary underpinning. Take the tick, for example. It's hardwired to move toward the light, so when it's in the tall grass and sees light through the leaves, it moves toward it. A respiring organism, say, a dog, walks by. It exhales CO_2. (Remember cellular respiration? Remember how your circulatory system takes CO_2 to your lungs and you exhale it?) The CO_2 triggers a fixed action pattern in the tick, which responds to this and another chemical stimulus by letting go of the grass. Sure, you'd rather imagine the tick, diabolically hanging off the branch, eying its target, thinking bloody thoughts. But it's really more mundane—some chemical signaling triggers the bloodsucking little animal to just let go. It lands on the dog. The dog is warm. Warmth triggers the tick to drill into the dog's skin and start feeding. Eww. It's gross. But it's also a really good example of how nature shapes innate responses to ensure an organism's survival.

Habituation and Conditioning

All of these innate responses require no learning. What is learning? It's a behavioral response to a specific previous experience. You may have heard that a smart person learns from his or her mistakes. Well, as it happens, the capacity to learn is a sign of higher-level cognitive functioning. Of course, even flatworms can learn, so we haven't climbed quite that high, yet.

> **Bio Bits**
>
> Flatworm school: in a test involving food dangled in water on a string, swimming flatworms, after randomly encountering the food, eventually learned that if they located the string, they could follow it to the tasty payoff.

The first level of learning is simple *habituation*. Take a sea anemone (careful; they're sting-y). If you poke it with a stick, it withdraws in an innate, protective reflex. But this sudden, frantic withdrawal takes a ton of energy. If you keep poking the poor anemone with a stick, it will soon stop the rapid, protective withdrawal and just sit there. If it had eyes, it would roll them at you in boredom. It has become habituated, learning that this stick, at least, is not a deadly predator intent on dinner. It's just some dumb, heartless human with a stick. The anemone has demonstrated simple learning.

On the flip side, *sensitization* is a process of learning in which the stimulus amplifies a response. The organism learns to be more sensitized or sensitive to the stimulus, rather than less so.

Another form of learning is trial-and-error learning, or operant conditioning. This is where that whole "learning from your mistakes" thing comes in. Fool me once, shame on you; fool me twice, shame on me, and all that. Smarter animals are good at this, and other animals may evolve to take advantage of it. A blue jay (a very smart bird) might, for example, try to eat a monarch butterfly. They're beautiful but toxic creatures, and within a short time of ingesting its winged prey, the blue jay hurls. It learns never again to eat a butterfly that looks like that. It's gotten a negative outcome from its behavior, and won't do it again. If the outcome had been positive—maybe a delightful feeling of fullness and abundant energy— then this reward would have triggered a repeat of the behavior.

def•i•ni•tion

Habituation is a decline in response to a repeated stimulus and is a modification of an innate response. **Sensitization** is the amplification of a response to a repeated stimulus.

It just so happens that another kind of butterfly, the viceroy, looks a lot like the monarch. It takes advantage of the ability of the blue jay to learn. By looking like a monarch, it benefits from the blue jay's trial-and-error experience and also may avoid being targeted for dinner.

Animals also exhibit associative learning, in which the organism associates a response to one stimulus with a completely separate stimulus. The classic example of associative learning is Pavlov's famous dog. These experiments also provide the seminal example of classical conditioning, so let's describe them here.

First, you need a dog. Then you need an unconditioned stimulus that elicits an unconditioned (unlearned) response. Food as the stimulus will bring on drooling, the unconditioned response. Now you need a conditioned stimulus, one that would only bring on the response if the organism learns to associate it with the unconditioned stimulus. In Pavlov's case, it was a bell.

Bio Bits

Ivan Pavlov (1849–1936) was a Russian physiologist whose interest in digestion led him to his discoveries about reflexive drooling in dogs. He was awarded the Nobel Prize in Physiology or Medicine in 1904 for his work.

Offer the two stimuli at the same time, proffering food while ringing a bell. The dog drools. Do this for awhile.

Now remove the food stimulus and only ring the bell. If the dog drools, it's demonstrating an unconditioned response to a conditioned stimulus, one that wouldn't normally produce drooling (unless the dog has some aberrant love for bells). The dog has been classically conditioned to drool to the bell. The dog has also exhibited associative learning, associating the bell, without even thinking, with its "there's food" response.

Imprinting: Mom's a Duck

Perhaps you've seen the pictures: an older man named Konrad Lorenz, walking down the road or wading through a pond, trailed by a string of waddling or paddling birds. Why are the babies following him? Is he the pied piper of water fowl? No. They just think he's their mother.

Nature has devised what ought to be a fairly foolproof mechanism for survival in some species. They hatch—usually, they're of the hatching kind—from their egg, and the first thing they see move, they decide that's Mom. From there on, they follow Mom wherever, into ponds, out of ponds, across the street. Because mom is a duck (or goose or some other fowl), she's got a lot of kids to lead around. This particular adaptation in her offspring, called imprinting, means she doesn't constantly have to count heads. They'll just be there, following her around. Or following famous ethologist Konrad Lorenz around.

> **Bio Basics**
>
> Imprinting is a kind of phase-sensitive learning, which refers to learning that can only happen during a particular window of development. Some people say, for example, that in humans, our sensitive phase for learning a moral groundwork is early childhood.

That generally works pretty well. After all, the world is full of ducks, so it's a success.

Insight Learning: A Stick Is Good for ...

Have you ever had a dog? Perhaps you've taken your dog walking on a leash, and he's gotten it wrapped around, say, a tree trunk. You're tired. You stand there, calling, "Come on, Dug. Come on! Just back up, unwind!" But Dug just pulls harder in the wrong direction. That's because Dug does not have much capacity for *insight learning*.

> **def•i•ni•tion**
>
> Insight learning is classically defined as "the sudden production of a new adaptive response not arrived at by trial behavior or as the solution of a problem by the sudden adaptive reorganization of experience." (W. H. Thorpe. 1956. *Learning and instinct in animals*.)

Species that show insight learning have the capacity to solve a problem without ever having encountered it before. Often, the smartest species in this regard are also social species: primates, corvids (crows, magpies, ravens), elephants, and dolphins. In fact, one crow gave us one of the most startling and revelatory examples of insight learning in recent memory. Confronted with an upright tube at the bottom of which sat a little metal basket of food, the crow devised a solution. She took a handy straight piece of wire, bent it around a nearby object to make a hook, and inserted that hook into the tube to grab the basket handle and pull up the meat.

This is really rather mindblowing. Not only did this New Caledonian crow exhibit the capacity to make a tool (so much for that being a "primate" thing) but she also solved a complex food-acquisition problem by doing so. She'd never encountered meat in a basket in an inaccessible tube before. Her solution is a true example of insight learning.

Cognition: Smart? How Smart?

Cognition has many meanings, depending on your discipline. Here, we're going to consider it as the ability of an animal's nervous system to perceive, store, process, and use information gathered by sensory receptors. If that sounds really biological, well … this is a biology book. The better an animal is at doing the above, the better their cognitive abilities are.

> **Bio Basics**
>
> Why do animals play? Play behavior is something we see only among animals we think of as being pretty smart. So bees don't play, but puppies and people do. Why? Play may be practice for adulthood: learning to ritualize aggressive interaction, obtain food, avoid predators, and practice parenting.

Cognitive abilities include giving attention (often to many different kinds of input: orange, large, quiet walker, smells like a tiger), being able to categorize (will eat me/won't eat me/might eat me), having memory (I remember that this is something that will eat me), and solving problems (this thing is trying to eat me—how do I get away?). "Higher-order" cognitive abilities include using complex communication such as language and having a conception of self. Animals that are considered from a human perspective to be good problem solvers are also considered to be cognitively superior. That includes, of course, ourselves.

Animal Behavior: It May Surprise You

This human-centric approach to understanding behavior may have misled us in some cases to believe certain animals lack specific capacities that they genuinely have. Take the bee, for instance. Just an insect, right? Probably not too bright in the grand scheme of things, right? Well, bees not only can detect their location spatially, they can return to their hives and communicate the exact location of an especially rich food zone through a series of dance moves that we affectionately call the "waggle dance." Other bees can understand this communication and follow its directions to the food source.

We also tend to think of ourselves as the only animals able to recognize self as self. Look in a mirror. Put your hand on your head. You know that's you, there, putting your hand on your head. A chimpanzee or a very young child might think that's another member of the species moving in the mirror and try to reach out and touch them. What about elephants? While the jury's still out, at least one elephant has shown the ability to recognize herself in the mirror, even understanding that a big white mark painted on her forehead doesn't naturally belong there.

Communication and Behavior

Members of a species usually must communicate with each other and often with other species. Communication covers many interactions, from mating to defense to camouflage to territoriality. What we're talking about here is animal communication, which often relies on aggressive interactions. Even mating sometimes looks more like fighting than the human perception of it as somehow romantic.

Hey, Baby! Mating

One early summer, my kids came rushing into the house, excitedly screaming that they'd found two snakes "fighting" outside. As a biologist, I knew right away that the snakes were likely not fighting, but mating. First of all, snakes, like most species, don't usually engage in deadly hand-to-hand or scale-to-scale combat. Second, mating in the animal kingdom often looks a lot more like fighting, or agonistic behavior, than anything else. Indeed, the two reptiles were mating. I decided not to disabuse my very young children of the notion that they were fighting, however. It wasn't that far off the mark anyway.

 Bio Basics _____

Agonistic behavior often is more about ritual, symbolism, and threat than causing actual harm. A classic example is the clash of two heavily antlered bull elk. They could likely kill each other, but usually they emerge physically unscathed, even though one has been proclaimed the winner and gets the girls.

Mating in the animal kingdom can involve many environmental inputs, from the chemical cues of pheromones to the visual cues of a primate's brightly colored behind. Some species, like the stickleback (a fish) or the crane (a bird), have elaborate mating rituals in which each potential partner must recognize the order of steps and participate appropriately. Of course, underlying these behaviors and visual changes are hormones.

The stickleback uses both color and movement in mating. The male's belly turns red when it's time to mate, and he uses it to scare off other males. He builds an elaborate nest of algae and kidney secretions (yes, kidney secretions), after which his belly turns blue. Female sticklebacks are apparently drawn to blue bellies and kidney-secretion nests, and a heavily gravid female will respond appropriately to these overtures by showing off her egg-enlarged belly. Each of these steps—the color, the nest, the belly demo—is required for sticklebacks to mate successfully.

Bio Basics _____

While we first discovered pheromones as chemical sex attractants in insects (silkworm moths), we've come to find that they are ubiquitous and diverse in the animal kingdom. Some studies have suggested that women can influence one another's menstrual cycles via pheromonal communication.

That's Mine! Aggression and Territoriality

You'll note that in the above example, the stickleback has staked his territory and defends it with his frightening red belly. Usually but not always, the males of the species are charged with territorial defense. Often, the defense or aggression and mating ability are coupled in some way—the oversized antlers of a bull elk come to mind. They help him both defeat rivals and keep his harem of cows. The same hormone that drives development of those antlers and that mating behavior also drives his aggression: testosterone. We are all, in the end, slaves to our endocrine systems.

Most animals and pretty much all vertebrates exhibit a great deal of territoriality and aggression. In fact, aggression is a defining feature of vertebrate behavior. Think about getting into an elevator that already has two people in it. Do you all huddle together in the corner? No. You each stake out your elevatorial territory. If someone were suddenly to try to huddle with you, in your space, you'd probably feel

Biohazard! _____

Don't walk away with the assumption that this territoriality and hormone business is solely an animal thing. Plants have hormones, too, including some that are a lot like animal hormones, and they also can be territorial in their way.

at least a bit aggressive—and likely totally creeped out, too. That's what violations of territoriality do to vertebrates. You can thank the hormones.

Often, chemical messages suffice. That's why dogs pee on trees, bears rub their backs on bark, and cats spray the furniture. There's many a reason to stake out and defend a territory: food, mating, rearing young, or having a safe place to sleep. Would you defend your home against threats for any one of those? That's because you're an animal, and that's your territory.

To Rear or Not to Rear: Parenting Behavior

Not all animals parent. Even in the same taxon, parenting routines can differ. Turtles, for example, lay eggs and vanish. Crocodiles, on the other hand, are pretty good parents. What triggers this urge to parent (or not)? One trigger is prolactin, first identified as the hormone that drives milk production in mammalian females. Another "parenting hormone" is oxytocin. In addition to driving those powerful uterine contractions and aiding in milk expression, oxytocin is also the "trust" hormone, counteracting the natural tendencies of vertebrates to want to beat each other up. This allows parents to care for their young.

Where's the Food? Foraging Behavior

Everybody's gotta eat. But what drives an animal out into the forest or desert, in the face of all danger, looking for food? Once again, hormones are in charge, directing the animal's desire to forage, or search for food. There are counterbalances to this risky behavior; for example, some individuals of a species may have a natural tendency to forage in more shielded areas, while others might have a preference for the open. The success of either depends on which their predators prefer. Either way, foraging is a must-take risk for survival. The key is the balance between the risk (of getting eaten) and the payoff (caloric acquisition, nutritional value).

Altruism: Real or Romance?

Altruism in biology is the exhibition of a behavior that reduces the acting individual's fitness but boosts that of another individual. What does that mean? Well, your fitness is a measure of how well your genes survive in the population. Obviously, if an assassin threatens someone near you with a gun and you leap in front of the bullet, getting killed in the process, you've sacrificed your own fitness for the other person, while increasing theirs. That was pretty altruistic of you.

Some scientists debate the true existence of altruism. After all, there is often some sort of recompense for a "selfless" act. If you help a stranger to your own peril and don't die, you've got all those back-patting endorphins (pleasure neurotransmitters) making you feel great about yourself. Say you feel so great, you go and reproduce that very night with your spouse. Looks like you might have enhanced your fitness after all.

The Least You Need to Know

- Behavior arises from a mix of genes and environment governed by the nervous and endocrine systems.

- Behavior encompasses what animals do in mating, parenting, foraging, fighting, and playing.

- Innate behaviors are hardwired, and while some are unchangeable fixed action patterns, others can be modified.

- The simplest learning is habituation, such as altering a defense response to a stimulus that causes no harm.

- Other forms of learning are operant conditioning (trial and error), associative learning (Pavlov's dog), and insight learning.

- True altruism is rare.

Chapter 26

Populations and Ecology

In This Chapter

- ◆ How populations distribute themselves
- ◆ Visualizing population structure
- ◆ Strategies for survival
- ◆ Limiting populations
- ◆ Carrying capacity

The population is the smallest biological unit that evolves. But how does this unit interact with its environment? What determines who survives and who doesn't, where they live and how long? Populations exhibit patterns of distribution and structure. They have the potential to grow exponentially but the environment often puts up resistance. Ensuring survival means adopting specific life strategies from birth to death.

Population Distribution

The individuals in a population will distribute themselves in patterns. Often, the driving force behind these distributions is interactions among the individuals or with other species, although abiotic (nonliving) influences also can determine distributions.

The number of members of a population in a defined area is the population's density. Take a square mile of a city. If skyscrapers cover the area and reach up 22 stories, each with several apartments, that square mile will be more population dense than a pasture in Iowa covering the same area but having a single farmhouse. How individuals in the population distribute themselves in the given area is their dispersion. As we soon see, dispersal tends to occur in patterns.

How Organisms Assort Themselves

When we go out into crowds, how do we assort ourselves? Do we all cluster in a huge pile on one side of the available space? No. We tend to stake out our little square foot or so of territory and then defend it from all comers (unless, of course, they're bringing food). The next time you go to a big event—a show or fair or festival—watch how the crowd behaves. Without discussion, without any obvious interaction, people disperse and assort themselves in ways that usually are unconsciously designed to reduce contact, friction, and agonistic interactions. Try it. You'll have a new take on being part of the crowd.

Population Distribution Patterns

Population distribution patterns fall into three basic categories. Organisms may clump together into little groups, assort themselves in uniform distances from one another, or appear to have no pattern of distribution at all.

Clumping usually arises from positive interactions among individuals in the clump. Go back to our crowd. That group of girls walking along together, giggling? They're a clump, interacting positively. Clumps often are under the influence of abiotic factors, or what biologists call a "patchy habitat." An example might be organisms clumped together under the sparsely available shade of large rocks, hiding from a hot Texas sun.

Clumps are useful in that the clumpees may be able to fight off predators more effectively, although as a smelly, visually obvious group, they may also attract them more easily. Another advantage of the clump is the "it takes a village" concept: the young have more role models in a clumped dispersion.

So why go uniform? Usually, these patterns arise from negative interactions among the individuals. Think of the elevator scenario. If you tried to clump with people on an elevator, they'd be pretty annoyed. We tend to take on a uniform pattern of equal-as-possible distance from the people around us because, well, we're vertebrates and we tend to be naturally aggressive and territorial. Just check out housing developments and their uniform distribution. There's a reason for that.

Random in nature is usually not truly random. This kind of distribution might exist only as a transient condition after some sort of catastrophe or unusual event like Woodstock. The wind blowing some seeds around might produce a transiently random distribution of the seeds. But nature would soon set that to rights, with the plants growing either in a mutually beneficial clump or engaging in a competition that produces a uniform distribution.

Population Dynamics and Growth

Populations can change over time—remember? That's evolution. But even without considering a change in allele frequencies, the makeup or nature of a population also can change. At one point, there may be far more reproducing individuals than older ones. At another point, the population may be running short on individuals in their reproductive prime. In the first scenario, if females are in sufficient supply, the population may be growing because its births may outnumber its deaths. In the latter scenario, the population may be shrinking because death outpaces its births.

Bio Bits
While the global human population continues to grow rapidly, the populations of many countries, such as some in Europe, are shrinking.

Deaths and births are the major factors in population growth and dynamics. Other factors, as we shall see, are those that influence deaths and births. But as you can probably intuit, a population with a large reproductive base is likely going to grow quite rapidly compared to one that has a diminishing reproductive base.

Other factors that influence the rate of growth are the simple (or not so simple) acts of leaving the population (decreasing it) or joining the population (increasing it). Thus, deaths and emigration shrink the group, while births and immigration grow it.

Demographics and Life Tables

Demography addresses information about the population like its birth rates, death rates, and ages of the individuals. Often, these data can be presented in tables that break down the population into age groups and then monitor how many individuals of each sex die in each age group in a specific period of time.

Survivorship Curves: K vs. r

While tabular information is useful, sometimes a visual representation tells a complete story. In fact, the information can be so complete that we can classify organisms based on what kind of curve their demographic data yield. Roughly sketched out, there are three types of survivorship in the natural world, usefully designated I, II, and III.

A Type I survivor would be an organism like the chimpanzee. They have few offspring and provide a lot of parental care, and a relatively large percentage of offspring survive to adulthood.

Bio Bits

Ever wonder how researchers figure out life expectancy for a specific age group? A crystal ball? No. With enough data on life tables, people can draw conclusions about how long a specific age group can anticipate staying alive.

On the other end of the survivor curve spectrum are the Type III survivors. Their life history strategy involves flooding the world with offspring, providing little to no parental care, and having a small percentage of the offspring survive. That little sea turtle hatchling that we cheer on as it struggles toward the ocean, ushered along by jubilant humans? It's probably gonna die. Survival can be 1 percent or even less for some Type III survivors.

Type II falls, as you may have guessed, in between I and III. There may be parental care. Females likely produce a middling number of offspring—not thousands at once, but not only one or two, either—and a middling number of them survive. The usual example of this kind of survivorship organism is the squirrel.

Biohazard!

While the term "life history strategy" suggests some kind of active planning on the part of a species, don't be fooled. Nature, not the organism or population, maps out the strategy through selection based on the environmental context.

We also can classify species based on a generalized life history strategy into r- or K-selected organisms. In this construct, r refers to "rate of increase," and these organisms rely on a lot of births to overwhelm their death rate, often attributable to early loss. Thus, their strategy is a constant rate of increase through massive influx of offspring, with the expectation that many of them will die.

The *K* in K-selected refers to how many individuals of this species their environment can support, or carry. I guess that should be "karry" because the *K* refers to *carrying capacity*. Because of their longevity and other characteristics, these kinds of organisms fall under control from factors that determine how many of them their environment can withstand. The table contrasts the r- and the K-selected attributes.

def•i•ni•tion

Carrying capacity, given as *K*, is the largest number of individuals in a population that the environment can support.

Comparison of Some Typical Traits of r-selected vs. K-selected Organisms

r-selected	K-selected
Opportunistic	Maybe more specialized
Many offspring	Few offspring
One-time reproduction	Periodic reproduction
Type III survivors	Type I or II
Small offspring/adults	Often larger mammals/birds
No parenting	Long growth period/maternal care
Not so adaptable to disturbances	May do okay with disturbances
Often short lived	Longest life spans

Tradeoffs: Parent One or Ignore Many?

This big decision about whether or not to have few offspring and parent a lot or have many and parent not at all is what we call a tradeoff. Most decisions about reproduction involve tradeoffs from the minute conception occurs, or even before. Vaccines carry risks, but the tradeoff for taking the risk is warding off the even greater risk from the diseases they prevent. Getting into a car carries risks, but the tradeoff is that we get places in the car a lot faster. Life is all about these kinds of tradeoffs—does the good outweigh the bad of this decision?

In the animal world beyond our own, the cost-benefit analysis has been one developed across millennia. The reason the sea turtle's strategy is to flood the oceans with offspring and provide no parental care is that it's worked for a really long time. Yes, most hatchlings die. Whether it's because they intuitively hide better than others, eat better than others, swim faster than others, or all of these and more, the ones that don't die survive to reproduce and continue this Type III strategy. The tradeoff continues to work to their benefit.

We have our own set of tradeoffs. One may be the tradeoff associated with growing our enormous brains. Human infants emerge more helpless and underdeveloped than almost any other placental mammal. We've got to stay in just long enough to get good lungs, but we can't stay in there too long growing those big heads because if they get too big, we can't get out. There's a fine line between what the pelvic opening can withstand and where we've arrived in terms of cortical development. One tradeoff is that when we're born, we still have several months ahead of what is essentially a continuation of fetal development. That requires some serious parental care. We've traded the risk of early emergence and the requirement of parental care for the benefit of having enormous brains.

Limiting Populations

With these strategies sculpted over millions of years for a given species, why is it that their success doesn't lead to every species simply swarming over the entire planet? Well, no species is an island. Every population runs into a little limitation we call the carrying capacity, which, as you just learned, is what the environment can bear.

Carrying Capacity: What the Environment Can Bear

I read recently about a woman in the *Guinness Book of World Records* who gave birth to an astonishing 69 children, including multiple instances of multiple births. Regular women who don't produce triplets with every other pregnancy might have the capacity to have, say, 20 children in a lifetime. But most women don't; in fact, right now, the average woman in the United States has 2.1 children (insert "how do you have 0.1 children" joke here). But that ability, that potential, for most human females to have that many offspring actually has a special name: it's our *biotic potential*.

def•i•ni•tion

Biotic potential is the maximum reproductive output that could be achieved under the best-possible environmental conditions.

We tend not to fulfill our biotic potential for many reasons. While with humans, cultural influence has a great deal of say in these things, in the natural world, biotic potential suffers under the crushing weight of environmental resistance. Thanks to this resistance, most organisms don't completely fulfill their biotic potential. That's really a good thing in the long run.

Environmental Resistance: Limitations on Growth

You may be able to think quickly of a few factors that limit a population's biotic potential. Water is one that comes to mind. Others include food, habitat, and other resources, and factors like disease, predators, or parasites. Those factors that are most effective in limiting a population when it's dense are called density-dependent factors. Predation is a good example of a density-dependent factor; densely packed populations make good targets for predators, while sparsely dispersed populations don't make for such great hunting. There's a reason lions tend to stick to herd animals for food.

Factors that limit a population regardless of its density are called density-independent factors. These are more difficult to pin down, but they can include the results of human activity, such as pollution, or natural catastrophes, like a tsunami.

Understanding the Graph of Effects

Most books that have chapters on population ecology include the classic graph depicting carrying capacity and a population at equilibrium. Not to be outdone and because it really does tie together most of the concepts of this chapter, I've included one on the following page.

As the graph indicates, when a population is in a growth phase, it's in the process of fulfilling its biotic potential. That's the exponential, climbing part of the curve on the graph. But it's eventually going to meet the environment's incapacity to handle more, because of resources and other limitations. That's the environmental resistance. Of course, this resistance has been active all along, but at a certain critical mass, it really kicks in.

So you may be wondering: does a population ever exceed this capacity, and if so, what happens? The answer is, Yes. And "Crash!" Some populations actually hum along in boom-and-bust cycles in which they far overshoot carrying capacity for a brief, shining moment, only to crash to the bottom just as rapidly. They then, with much reproductive effort and some recovery of the environment, cycle back up again. Lemmings are a well-known example of this kind of boom-and-bust cycling. Locusts, immortalized as one of the biblical plagues, also use boom and bust as a life-history strategy.

Population tends to hover right around the balance between its biotic potential and the counterweight of environmental resistance. This natural calculus produces the ultimate outcome: the carrying capacity for that population.

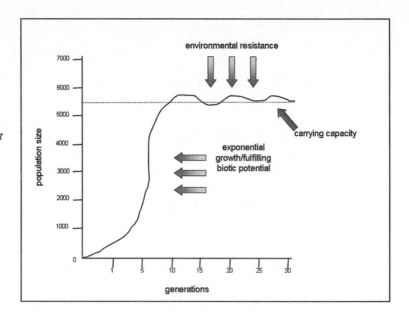

Bio Basics

A famous example of a boom and crash happened on St. Matthew Island off of Alaska's coast. In 1944, people left 29 reindeer on the island, dining sumptuously on a rich abundance of lichens. By 1963, the population had exploded to 6,000. But the lichens had run out, and in 1964, the reindeer population had bottomed out to only 50.

The Future of the Human Population: It's Big

Speaking of booms: when I was an undergraduate, there were about 5 billion people on Earth. Now the number's close to 7 billion. That's 2 billion people in the last 20 years alone. Where is the human population headed? Are we doing fine, humming along toward the planet's carrying capacity for our species? Or have we passed that mark long ago and are headed for a bust of lemming- or locustlike proportions?

The answer? No one really knows. Sorry. But projections are a wee bit terrifying. By 2040, we might be up to 9 billion or more. The good news—if you're not fond of crowds—is that growth has leveled off and is expected to stay that way. The bad news? There are about 134 million births a year, which isn't expected to increase, but deaths are at about 57 million a year. As we've just learned, as long as births outpace death, we have exponential growth. While we can't predict, necessarily, where we are on our carrying capacity graph, one thing is certain: we're still soaring upward.

The Least You Need to Know

♦ Populations may be distributed because of biotic or abiotic factors and fall into clumps or a uniform distribution.

♦ In a growing population, births exceed deaths and immigration outpaces emigration.

♦ K-selected species produce few young and engage in parental care, with the tradeoff of more young surviving.

♦ r-selected have many young and little parental care with the tradeoff of some of the many surviving.

♦ Carrying capacity is what the environment can bear, and biotic potential drives a population's growth.

♦ The human population has been growing rapidly for decades, and no one knows the planet's carrying capacity for us.

Chapter 27

Community Ecology

In This Chapter

- ◆ Defining community
- ◆ Finding a niche
- ◆ Community competition and compromise
- ◆ Co-evolutionary relationships
- ◆ Succession and building communities

Community ecology is the way that species interact with each other to build and maintain a community, an interacting aggregate of different species in a geographical area. These interactions can be friendly or indifferent or deadly, and they influence the evolutionary paths of every species involved in them.

Defining the Community

I keep saying it: no species is an island (with one possible exception). Thus, we usually find different species living near one another and interacting in communities. As in any community, some species get along, some compete fiercely, and some take advantage of others, to the latters' detriment.

It's not much different from human societies, except that communities are based on interspecies interactions, while our human communities rise and fall because of intraspecies interactions.

Biomass is the total organic matter making up the organisms in a given area.

Also as in human societies, some contributors to the community structure carry more weight than others. These keystone species can be a lynchpin linking community interactions. Another key factor in the community is the dominant species, usually the one that exists in greatest abundance, either in terms of its *biomass* or sheer numbers of individuals, or both.

On land, we often designate communities based on the dominant plant species in the area. Thus, a community in northern California might be described as a mixed-shrub-tanoak-Douglas fir community. Where the eye can't see, billions of bacteria lurk in communities that consist of assemblages of different species. And in the oceans, communities take their designations from the depth at which they occur. Organisms that occupy the lowest layer of a body of water, the benthos, which includes the sediment, make up a benthic community.

Niches: Fundamental vs. Realized

Communities can be a lot like high school, with niches and cliques and everyone assigned, theoretically, to a specific level in the pecking order. Everyone has a role: jock, cheerleader, band geek, Goth baby. In community ecology, species also have their specific roles, their niches.

What determines a species' niche? Many factors can be involved, including who eats it, what it eats, where it lives, how it breeds, whether it's r-selected or K-selected, a Type I, II, or III survivor … all of these influence its role in the community. Obviously, changes in the community structure or interactions can lead to changes in job function, just as changes in conditions in the human workplace can lead to changes in the roles we play. If you're a key r-selected species in your community and a new r-selected species floods the community with offspring, your role as the dominant r-selected species may change.

Biohazard!

You may be tempted to think of a niche as a physical place to occupy, but it's really the species' role—what it does as its job in the community.

Competitive Exclusion

This competition between two similar species, like that between these r-selected superproducers, can lead to a problem. In ecology, no two species can occupy exactly the same niche. In such a head-to-head competition, to use the environment in exactly the same way, to serve in exactly the same role, something has to give. This principle, called the competitive exclusion principle, predicts that this direct competition for exactly the same role will lead to extinction of one of the competitors unless they compromise. As in a Broadway play, in a community, it's only one player per role, and there are only so many roles to go around.

Too Much Alike

Usually, this kind of competition arises only between species that are already very similar. You wouldn't expect, for example, there to be a threat of competitive exclusion between an oak species and the wombat because these organisms simply do not use the environment in the same way. But if the competition were instead between two very similar species of oak, both with the same resource requirements and use, one of them would likely fade out of existence in that particular area.

A Mechanism of Extinction

The empirical example of competitive exclusion you find in books most often is one we can't even see: a laboratory competition between two very similar species of protists (single-celled eukaryotes). When the organisms were allowed to grow in separate flasks with a stable food supply, the population growth followed the classic carrying capacity curve for both species. Their numbers climbed to a certain height and then plateaued, held at carrying capacity by environmental resistance in the form of the stable, but limited, food supply.

But when both species were placed together in the same flask and again supplied with a stable food source, something rather surprising happened. You might think that some members of each species got food, while others were not so successful, so that each population perhaps underwent a reduction but still survived. But the actual outcome was different: One species of protist actually became extinct (within the flask), driven out of existence in this life-or-death competition with its overly similar competitor. The second species evidently had some competitive edge in this battle to use the environment (the food) in exactly the same way.

Resource Partitioning

You may be thinking, "But what about all of those species of cichlid in Lake Malawi in East Africa?" If that's not what you're thinking, I'll explain. Cichlids are fish, and if you've ever perused the tanks in a pet store, you've probably seen them. In this large East African lake, there are about 280 documented cichlid species, but up to 500 species may actually occupy its waters. It's a big lake, but there still are hundreds of what are often very similar species of these fish. How is it that these fish species, some of which are so alike that it takes specialists to tell them apart, manage to live side by side or fin to fin, often occupying what must be pretty similar niches?

Compromise Saves a Species

The answer is compromise. If you've ever had to share a room with a roommate or a sibling, you may have some idea of what kind of compromise this might entail. The room probably had resources—beds, desks, lights, a sink, maybe a bathroom. In some cases, you could simply divide the resources in half: each of you gets a bed, a desk, a lamp. But what about a resource like the bathroom? You couldn't both use it at once. How do you compromise? One possible solution might be a simple matter of timing: you just use the bathroom at different times, or one uses the sink while the other uses the shower. Cichlid species in Lake Malawi make similar fine distinctions about resource use in their bid to live and let live.

That bathroom compromise isn't much different from the unspoken agreements that species reach to avoid the intense competition that could lead to extinction. We call these unspoken agreements resource partitioning. It's the divvying up of resource use as a result of interspecific competition.

Divvying Up the Niches

The first thing to understand is that species often don't get to use resources as fully as nature might allow under unlimited resource conditions. The presence of other species and their resource requirements tends to interfere. Back to our roommate example: you might have the potential to have the entire bathroom to yourself at all times, but circumstances—dorm room affordability, for example—have reduced that potential to a different reality. These circumstances have forced you from fully occupying your *fundamental niche* and limited you to occupying only your *realized niche*.

The potential parallels to human interactions are legion. In the workplace, you may have tons of potential—perhaps you're fully capable of running the place—but there's competition, so even if your fundamental niche is the corner office, your realized niche is a doorless cubicle, thanks to workplace competition. Cubicles are really the perfect example—literally—of human resource partitioning.

def•i•ni•tion

A **fundamental niche** is the role that a species might play in theory, in the absence of competition. The **realized niche** is the reality: the part of the environment they really do end up using.

Scientists have made some pretty cool observations of resource partitioning in the natural world, too. A famous pair of partitioners are two barnacle species off the Scottish coast. When both species are present along the rocky coast, one occupies an area higher on the rocks, while the other glues itself to rocks closer to the surf. But when researchers moved the lower-sticking species from an area, the higher-up species immediately began occupying the newly available space. Thus, the higher-living barnacle species had a fundamental niche that was much larger than the one it actually realizes under natural conditions. When its interspecies competition disappeared, it expanded into its fundamental niche. When both were present, the compromise of dividing the rocky shore into high and low environments kept either species from becoming extinct.

Sometimes, this compromise over niches can lead to selection for differences in characteristics related to success in the realized niche. When a feature that two species share becomes different when they compete, we call that effect character displacement. We can tell the interspecies competition displaces a characteristic because in the absence of the competition, the characteristic remains unchanged. An obvious visual example is size. Individuals from two closely related salamander species in the United States are always about the same size if they live apart from each other. But wherever the geographical areas they occupy overlap, individuals from one species are clearly larger than members of the other species. Close and intense competition between the two species resulted in displacement of the character of body size.

Predation Interactions

Interactions between species in a community aren't only about fighting each other for resource use. Any time you've seen a lion take out a gazelle on one of those nature shows, you've seen interspecies interactions that aren't about competition. Obviously, they're really about using resources; in this case, how the predator uses the prey.

Defenses: Camouflage and Chemicals

And not all potential prey species fit the image of the helpless, quavering, toothpick-legged gazelle leaping in terror about the savannah with a merciless lioness in pursuit. The prey engages in interaction with the predator quite as directly as the predator targets its prey.

Even we engage in this kind of interaction with prey when we hunt. Camouflage, which in biology we call cryptic coloration, works both ways. A predator may use it to conceal its approach to the prey, and prey may use it to conceal themselves from the predator. The beautiful markings of snakes often serve the former purpose, while the white spots on a fawn's back help conceal it as it lies in the dappled light under a shrub.

def•i•ni•tion

Aposematic coloration designates the bright colors that some organisms present as a warning of their dangerous toxicity or other defense.

Then there's the opposite of cryptic, when organisms adopt the bright colors of nature as a warning that says, "I can kill you with my deadly toxins." Sometimes, the bright colors of *aposematic coloration* tell the truth: the poison dart frogs of South America, with their electric blues and greens and yellows, really truly can kill you. Other times, it's a ruse, as when the harmless milk snake manifests the bright red-yellow-black coloration of the venomous coral snake.

Mimicry: I Only Look Scary

This imitative gambit on the part of the milk snake is called mimicry. We've divided this coloration mimicry into two types, both named after the fellas who identified them: Batesian mimicry and Müllerian mimicry. If a nonvenomous or nonthreatening organism takes on the look of a deadly species, it's engaging in Batesian mimicry. In this kind of mimicry, the benign species picks up benefits by looking malignant, even in the absence of any real power to cause harm. The beautiful viceroy butterfly, which looks uncannily like its more toxic cousin, the monarch, reaps benefits from the similarity, especially from potential predators that have already experienced the nastiness of trying to eat a monarch butterfly.

Bio Basics

Batesian mimicry gets its name from the man who described it, the naturalist Henry Walter Bates, and my mnemonic for it has always been Batesian = benign (harmless). Biologist Fritz Müller gave his name to Müllerian mimicry, which I remember as Müllerian = malignant (harmful).

Müllerian mimicry occurs when the imitator is as harmful as the imitatee. The similar yellow-and-black coloring that many stinging insects share is an example of this kind of mutually advantageous aposematic coloration. Unlike Batesian mimicry, in which the nondangerous species doesn't add any advantage for the mimicked, Müllerian mimicry helps out both species involved. Any predator trying to eat either species learns a lesson.

Co-Evolution

These lessons learned between predator and prey are just one example of the constant process of co-evolution that takes place in a community. The species interacting in a community often have been doing so for millennia, changing and adapting with each other in a fine-tuned give-and-take of evolutionary change. The pollen producers and the pollinators. Insects and insect eaters. Ants and their fungi. People and their intestinal bacteria. Like old married couples, members of a community get used to each other, grow to understand and adapt to one another's quirks, recognize changes, and react.

This entire process of fine-tuning community relationships can implode, however, if somebody new suddenly shows up. While nature has busily made introductions of new species on her own throughout the existence of life, humans are responsible for the major debacles of recent memory. We brought cane toads to Australia to eat sugar cane beetles. The huge, toxic toads did nothing to the beetle population, but having no predators, they did multiply like crazy and continue to pose a real threat to the ecosystem and domestic pets.

Some introductions have been purposeful, like the cane toad, while others have been accidents. The South American fire ants are marching their way across the United States, stinging people without mercy and decimating native species on their way. With no natural predators, they can eat and build and move on with impunity. Boat propellers have spread *Hydrilla* all over the North American continent, choking lakes and water supplies with a tropical plant from Africa that also has no natural predators in its new home. Displaced from its own community with natural predators, introduced into a place where nothing gets in its way, this and other introduced species threaten the balance of any new ecosystem they encounter.

Commensalism: Only One Benefits

The interactions among species in communities can take on different characteristics and can be intense when two species are in symbiosis (living together). Sometimes, this intimate interaction between two species benefits both. Sometimes, only one benefits, while the other is unharmed. In still other situations, one benefits while the other suffers terrible harm.

In the form of symbiosis known as commensalism, one of the two species in the interaction experiences nothing either way, while the other benefits. Scientists argue over whether this kind of scenario can ever truly exist—surely, in every interaction, both parties involved must experience *some* kind of effect. At any rate, commensalism still exists as a concept in biology, and some examples include hitching a ride on another organism, as barnacles do on whales, or living on another organism without harming it, like an owl nesting in an existing hole in a tree. You may already have thought of some ways in which these interactions confer a negative or a benefit on either party, which would render them not truly commensal.

> **Biohazard!**
>
> Don't make the mistake of thinking that symbiosis always means friendly. It just means that the species live together. As with any roommates, the relationship could be warm, cool, or downright hostile.

Mutualism: A Mutually Beneficial Relationship

In the mutually beneficial interspecies interaction known as mutualism, both species involved reap benefits from the experience. In the case of leafcutter ants and the fungus they cultivate, the reaping is literal. These ants, as their common name implies, cut up leaves. But they don't eat the leaf bits. Instead, they are agricultural geniuses, mulching the leaves as food for their crops of fungi.

In turn, the ants eat parts of the fungi. This relationship benefits the ants because it feeds them, and it benefits the fungi because the ants care for and propagate the species. It truly is one of the most fascinating and fine-tuned examples of co-evolution, but there are many more. I'd recommend googling "orchids" for some unbelievable examples, one of which involves the orchid mimicking the presumably attractive female hindquarters of a pollinating species. In other words, it tricks males of the insect species into trying to mate with it and offloads its pollen in the process.

Parasitism: Bloodsuckers and Other Freeloaders

What is a parasite? There are a few ways to define one, but the definition I consider most accurate goes like this: one species uses a second species for some purpose, harming the second species while benefiting itself. Some parasites may set up shop on the outside of the organism and destroy it from the outside in. We're familiar with ectoparasites like this, such as ticks, which suck the host's blood. As you're likely aware, having something suck out your blood is harmful. Further harm can come if that something infects you with a disease-causing organism while it's dining on you.

Endoparasites cause their harm from the inside. The tapeworm is an endoparasite that lives in the intestines and can cause digestive upset. Maria Callas, the famous opera singer, is rumored to have deliberately infected herself with a tapeworm to lose weight, but she really came by it the usual way, by accidentally ingesting the parasite larvae in her food. Just to reinforce the "harm the host" aspect of the term parasite, let me urge that tapeworm consumption is not a healthy or safe way to lose weight. In fact, weight loss is not a typical sign of tapeworm infection.

> **Bio Bits**
>
> In a relationship that is definitely harmful to the host, a species of hairworm, as part of its life cycle, hijacks a grasshopper's nervous system and forces the zombie grasshopper to commit suicide by jumping into the water. The mature hairworm emerges from the dying grasshopper's anus and proceeds with the aquatic portion of its life cycle.

Succession

How does nature build a community? The process of the co-evolution of a community of species from its beginnings to its peak is called succession because it involves a gradual succession of species interacting with one another. And before there is any life in an area, there is bare rock. Nature literally starts the community building from the ground up.

Primary: Bare Rock to Woods

In the beginning, there is bare rock. Then microorganisms might move in, respiring and metabolizing and generally rocking out (ahem), until their activity breaks down some of that rock (gotta get phosphorus from somewhere!). The broken-down rock is now called soil. Windblown spores from lichens and mosses may also have weathered away at the rock. These colonizers have done what colonizers always do: they moved in, shook things up, and paved the way for others to follow.

With the freshly available soil, other organisms can make their move into the area, succeeding the colonizers. This next wave of succession might involve seed-bearing plants, the grasses and woody plants that lay the groundwork for other species to move in. Eventually, often over millennia, the succession continues until the dominant plant community that defines the area takes its place.

This process of going from bare rock to the dominant plant community is called primary succession. It's how any original community first arose. As you may have noticed, colonizers start things off. The way an area looks when succession begins—including the species that reside there—may be quite different from how things look at the end. Species replace species, colonizers may disappear, bowing out and giving up their places or the new niches they've created to the next wave in the succession.

Secondary: Farm to Forest

Secondary succession differs from the primary version in two ways: First, it doesn't start on bare rock but instead on already available soil, usually in an existing community. Second, it typically takes much less time to reach the climax community, that defining plant community that will give it its name.

If a volcano blows its top or a fire devastates a forest or a farmer gives up his fields, leaving behind denuded soil surrounded by the existing community, secondary succession will gradually fill in those gaps, eventually restoring the area, step by step, to some version of its original self.

The Least You Need to Know

- Each species in a community has a niche, or a role in the community.

- Similar species may compete for a niche and must engage in resource partitioning or risk competitive exclusion.

- Predator-prey interactions can involve camouflage and Batesian or Müllerian mimicry.

- In Batesian mimicry, a benign species mimicks a dangerous one, while in Müllerian, the mimic is harmful, too.

- Symbiotic interactions between species can be mutually beneficial or harmful or helpful only on one side.

- Communities arise in a process called succession, which can begin on bare rock (primary) or in soil (secondary).

28

Ecology and the Biosphere

In This Chapter

- ◆ Nutrient cycles
- ◆ Biomes of the world
- ◆ Aquatic ecosystems
- ◆ Energy flow and food chains
- ◆ About diversity and extinction

We started with atoms and we're ending with the biosphere, an interacting collection of ecosystems that cycle nutrients and transform energy and take shape thanks to climatic factors like precipitation and temperature. Traveling from small to big, we now gain an understanding of big—at least in earthly terms—by reverting to small. Yep, time to remember some things about the fundamental molecules and elements of life: water, carbon, nitrogen, and phosphorus. From atoms to Earth, it all comes together in interwoven webs we call ecosystems, that combined together make up our island biosphere.

The Water Cycle: Land to Sea to Sky

Nature is the great recycler. Animated films tell us about the circle of life, and it's true that the atoms that make us return to the dust to be taken up and made into something else. The nature of nature is change, sure, but it's also the nature of nature to cycle.

Almost any circle of life relies on water. Photosynthesis cannot happen without water molecules to split. And cellular respiration ends with the formation of a water molecule. Where does that water come from? Where does it go?

As you may be aware, water falls from the sky as precipitation, rain, snow, sleet, hail—all that stuff the mail carrier has to battle to get you your mail. Once it hits land, this water percolates through the soil into the groundwater stores or runs off into lakes, streams, and rivers. Some of it almost immediately returns upward thanks to solar heating. Animals and plants take up the water and discard some of it; plants through specialized cells in their leaves in a process called transpiration, animals through waste disposal. Once it accumulates enough skyward via evaporation, it can again fall as rain, continuing the cycle. Wind plays a role in moving water around, both in the earth's oceans and in pushing clouds across the skies.

The Carbon Cycle: Sinks and Sources

If you read Chapters 7 and 8, you've already encountered part of the carbon cycle. Plants fix carbon from carbon dioxide, build sugars that other organisms consume, the other organisms break down the sugars in cellular respiration and give off carbon dioxide in the process, from which plants then fix carbon.

Bio Basics

Whatever soaks up a nutrient is a sink, while anything that introduces a nutrient into the cycle is a source. Forests, for example, are typically considered to be carbon sinks because the trees soak up the carbon dioxide.

All living things require carbon, and it's a primary component of every major molecule of life. Thus, we need a lot of it. The earth has a few reservoirs of abundant carbon, including fossil fuels, living things, oceans, and the atmosphere. It exists dissolved in the oceans and as CO_2 in the atmosphere. While limestone is actually the largest carbon reservoir, life gets little use from this carbon because it's not released very quickly.

The Carbon Cycle and Global Warming

Human activity has affected all of the nutrient/chemical cycles we talk about in this chapter, but probably none has gotten as much press as the carbon cycle. That reservoir of carbon in fossil fuels is carbon packed away for a rainy day, something that on nature's timetable wouldn't be used for geological ages. Humans, not really caring about nature's timetable, have a different kind of requirement: gasoline-fueled engines. We extract the carbon prematurely from these reservoirs, turning them into sources, and burn the fuels, releasing the carbon into the atmosphere and doing strange things to nature's balance in this cycle. Instead of a drip-drip of carbon from this reservoir, we've released a deluge. And that deluge yields a warm thick blanket of carbon dioxide in our atmosphere.

The result is global climate change. The vast majority of scientists involved in such studies agree that the earth's climate is changing rapidly and that human activity is largely responsible.

> **Biohazard!**
>
> Don't confuse weather and climate. Some people experience an unusually cold day and think, "Hmmm … not much to this global warming." Weather is what's going on outside right now. Climate is an average of the weather in a geographical area over a long time period.

Ozone: Good and Bad

Where I live, when we have a hazy day with a lot of pollutants in the air, our local authorities call an "Ozone Action Day." Ozone (O_3) gets a bad rap because down near the ground, it's harmful. It's oxygen gas, O_2, gone rogue and transformed into O_3 by interacting with exhaust gases like nitric oxide. O_3 harms the respiratory tract and is thus bad down here.

But we also wouldn't live here without it. In the upper atmosphere, ozone forms a protective blanket against the pounding of ultraviolet light the earth gets from the Sun. In fact, it's so important that a growing hole in this layer was cause for great alarm for many years. The banning of chlorofluorocarbons (CFCs), a component of hairsprays among other things, helped shrink the hole. The CFCs were pretty much chewing the ozone to bits up there.

The Nitrogen Cycle: Prokaryotes Help Out

Like carbon and water, nitrogen is another chemical we just can't live without. Luckily, we've got lots of living things that keep the cycle moving so we can all stay alive.

Most of our nitrogen is in the atmosphere as nitrogen gas, although other more limited reservoirs are the oceans, soil, and, of course, all living things. Nitrogen can make its circuit in a number of different ways, many of them involving bacteria that access it from the atmosphere or other sources and modify it for plants to use or even giving off nitrogen gas as part of anaerobic metabolism.

The Phosphorus Cycle: Nutrient from a Stone

If you think about the molecules involved in life's processes that contain phosphorus, you realize its importance: DNA, ATP, RNA, phospholipids—all of these use phosphorus. As we've mentioned before, it's also the least available nutrient that's needed most by every living thing.

Luckily, there are a lot of rocks around. The major phosphorus reservoir is sedimentary rock, and our ability to access this phosphorus relates to the breakdown of this rock into soil. Once again, oceans and living things also are reservoirs of phosphorous—the one naturally home to much of the planet's biomass and the other the very thing that requires phosphorus in the first place.

Usually, phosphorus gets to other organisms via plants or organisms at the bottom of the so-called "food chain." These organisms take it up from the soil or water (dissolved in the oceans), and then other things get their phosphorus by eating them. Phosphorous cycles back via rainfall (if it's evaporated in water) or by decomposition of organisms.

Terrestrial Biomes: Interactions That Create Them

Climate is a major determinant of what lives where. The two climatic factors that interact to create the large ecosystem categories we call biomes are temperature and precipitation. High temperatures and low precipitation result in the biome we call the desert, while low temperatures and low precipitation produce the biome we call the tundra. We characterize biomes based on what kinds of plants are the dominant species in that kind of ecosystem. A few examples of terrestrial biomes and their features are discussed next.

Rain Forest: Rich and Humid

If it rains all the time and the temperature is warm, you're likely to be in a tropical rain forest. These are some of the most species-rich places on Earth, where a single tree can be home to dozens of ant species, where trees seem to grow on other trees, where life is layered on life from the ground to the treetops. This biome hovers around the equator where temperatures remain relatively stable year round.

Savannah: World-Famous Fauna

The lion isn't king of the jungle, because the lion doesn't live in the jungle; it lives on the savannah. The savannah biome is low on rainfall most of the year, with a long dry season and short monsoons. Trees are sparse and grasses are abundant, especially after rainfall. The "charismatic fauna" of the earth congregate here, from lions and cheetahs and hyenas to the herbivores that keep things rolling, including wildebeest and antelope and, yes, termites. Like tropical forests, the savannah biome occurs around the equator, and thus temperatures remain relatively stable year round.

Desert: Alive by Night

Desert biomes arise from a combination of low precipitation and temperature extremes, either hot (for hot deserts) or cold (for cold deserts, naturally). Their key feature is low moisture, which results in sparse vegetation coverage and organisms adapted to water conservation and with abundant defenses (think rattlesnakes, cacti, scorpions, creosote bushes). In hot deserts, cooler nights bring out the hunters.

Grassland: A Delicate Balance

While the great rolling prairies of North America may be the quintessential grass-land, the world boasts a number of examples of this biome on every continent except Antarctica. With summer rains and cold dry winters, the dominant feature of grass-lands is, um, grass. Although rolling prairie may not have much to delight the sensory-seeking eye, the American grasslands once existed in a delicate balance in which huge bison herds aerated the soil along with networks of prairie dog burrows, where fire renewed the grasses, and the grasses supported the animals. Now much of this vast, complex ecosystem is farmland.

Deciduous Forest: Shifting with the Seasons

Deciduous trees lose their leaves in the fall, renewing them again in spring. Temperate broadleaf forests cover biomes where rainfall is fairly abundant and temperature varies with the seasons. These forests, which occur primarily in the northern hemisphere, are home to organisms that can adapt to the fluctuating seasons, including hibernators and migraters.

Evergreen Forest: A Green Desert

I confess that I'm geek enough to have a favorite biome, and this one is it. Another feature primarily of the northern hemisphere, these forests don't get rainfall as abundant as the deciduous forest enjoys, and temperatures can be more extreme, especially the lengthy cold of winter and spring. The key feature of these forests is, of course, the conifer—pine, fir, spruce—that can withstand periodic drought, shake off heavy snowfall, and renew in the frequent fires that hit these dry, tree-filled deserts. While these forests are home to a number of charismatic species, including bears and mountain lions, they also suffer from unwelcome, if natural, visitors like the pine beetle, which is currently decimating much of the coniferous forests of the American west.

Taiga: Home of the Tiny Trees

Sometimes considered a single biome together with the coniferous forest, the taiga occurs specifically in the northernmost climes of the world. Taiga I've seen consists of sparsely distributed conifers, often quite short in spite of being quite old, limited in their height by the brief growing seasons and brutal conditions of the higher latitudes. Temperatures are truly extreme, with brutal winter lows dozens of degrees below freezing to highs in the summer that approach 90°F/32°C. Precipitation is fairly sparse. Animals that live here have adaptations for extreme cold.

Tundra: A Struggle for Life

Once you've climbed in altitude or in latitude to the places where trees simply don't grow any more, you've reached the biome known as the tundra. Tundra comes in second in my favorite biomes list. (What are your top-five favorite biomes?) It's a delicate environment that in summer blooms with miniature flowers and can resound with the squeaky song of the marmot, or whistle pig, while in winter it lies buried in snow and brutalized by bitter winds. It rarely gets what you might call "warm," but the vegetation is abundant enough to attract and periodically maintain large herbivores such as elk and caribou. In the Arctic tundra, bears and wolves have also managed to find and keep a foothold.

Aquatic Ecosystems

Given that this is the blue planet, it should come as no surprise that the largest biomes on Earth are not actually terrestrial, but marine. The aquatic biomes together add up to the majority of the biosphere and are home to most of the earth's biomass.

Whether ocean (marine) or aquatic (freshwater), water ecosystems are divided into layers from the top to the bottom. The first factor in determining the zones is whether or not light can penetrate, which in turn determines the kind of organisms that live there (photosynthetic or not). In the photic zone, light is abundant, as are photosynthetic organisms. The benthic zone lies at the bottom and includes the sediment, and in between is the apohotic zone, where light penetration is not enough for photosynthesis. If you go deep enough, the benthic zone hits depths of thousands of meters, and we call this zone the abyssal zone. The nonbenthic layers of the water from the top to near the bottom are the pelagic zone, or open water.

A freshwater lake near the shore has a littoral zone, where the water is shallow and light penetrates. Further out from shore is the limnetic zone, the open water away from shore but still including the photic layer.

Marine: Diverse Shallows and Deep Mysteries

Marine zones are quite similar to freshwater zones except that the littoral zone can also be known as the intertidal zone because its water coverage changes with the tides. The limnetic zone of freshwater becomes the oceanic zone in marine lingo, and between the intertidal and oceanic zones lies the neritic zone, marked at the shore where low tide hits and usually ending where the continental shelf drops off, sloping to the ocean bottoms and sometimes into that mystery of mysteries, the marine abyss.

The shallows, with their abundant sunlight and photosynthetic organisms, can be veritable species free-for-alls, with tidal zones yielding easy access to sea stars, anemones, and urchins, and coral reefs with their famed and colorful wildlife. The open ocean, or pelagic zone, is home to the behemoths of the ocean, such as whales, and to the sometimes-microscopic creatures that support our planet, the zooplankton.

Freshwater: A Soggy Succession

Freshwater lakes can undergo a process of succession similar to that which we described for terrestrial communities. If nutrient concentrations become rich enough, the water can gradually become so choked with organic matter that it can even stop being a lake and instead becomes land. While this process, called eutrophication, occurs naturally, especially with seasonal lakes, we have triggered it ourselves through pollution from agriculture and sewers.

Energy Flow: Chains, Webs, and Pyramids

Have you heard of the food chain? How about the *food web?* (I know you've heard of the food pyramid because we talked about it in Chapter 22.)

def•i•ni•tion

A **food web** is an interconnected grouping of food chains. They are a more accurate representation of reality than a simple food chain because one animal may participate in more than one food chain.

While many people think more in terms of "little fish gets eaten by big fish gets eaten by bigger fish gets eaten by human" when they think of the food chain, the real flow that this metaphor is supposed to illustrate is the flow of energy in an ecosystem. While a web is by far a more accurate representation of how energy moves around in an ecosystem, the chain is simply an easier image for our human brains to grasp. And it happens to fit in with our conceptualization of levels.

Trophic Levels

The position that an organism occupies on the theoretical food chain is its trophic level. We start at the bottom with things that photosynthesize. These organisms—plants if they're terrestrial or near terrestrial, phytoplankton if marine—kick everything off by using the Sun's energy to build organic molecules. Then everything else pretty much eats them.

Bio Basics

A food chain is really a representation of the flow of energy in an ecosystem. These chains usually have only a handful of levels because with more than that, the energy would dissipate completely before the end of the chain.

We call the organisms on this first level the primary producers because they produce organic molecules first. A step up from them, we have the organisms that directly consume primary producers, so we call them primary consumers. Next up the chain are the secondary consumers, which eat primary consumers. Naturally, our next group are the tertiary consumers, and finally, at the pinnacle of the chain, relying on everyone else below, are the quaternary consumers.

Autotrophs: Producers for the Planet

Now think about all of this in terms of energy. At the base of this chain are the organisms that must capture all of the energy that will eventually make its way through the chain, supporting every organism on it all the way up to the quaternary consumers. Given this duty of the primary producers, it should come as no surprise that they make up most of the biomass on the planet. We've got to have a lot of them to make up the huge energy base that supports everyone else. Take into account the fact that with every transfer of energy up the chain, a vast majority of it is lost as heat in the transformation, and the burden on the primary producers becomes even greater.

 Bio Basics

Primary producers represent the largest portion of the planet's biomass.

Consumers: Heterotrophy at Many Levels

So what about all those consumers? The classic illustration is that in the terrestrial world, an insect of some sort serves as the primary consumer. It, in turn, falls prey to a secondary consumer, perhaps some kind of rodent. A snake, the tertiary consumer, snags the rodent (perhaps on a cold night hunt in a desert biome?), only to become prey itself to a hungry quaternary consumer, such as an eagle. With each move up the chain, the levels below it must support the consumers above. As you can see, being a carnivore, especially a quaternary carnivore, means having a huge energy requirement that relies on four other trophic levels to support.

Where do we fall? This is where the obvious awkwardness of the food chain becomes apparent: where do omnivores fit in? We can be primary consumers because we eat plants. We can be secondary consumers because we eat herbivores, such as insects or cows. We can even be tertiary consumers if we eat other omnivores, such as pigs. What about quaternary consumers? Sure. In other words, either the food chain isn't quite right for us, or we fall into every level of the food chain except the primary producer level. Note also that if we stick to the primary consumer level, we use a lot less biomass to obtain the energy we need. By consuming at the quaternary level, we're taking up energy after several transformations and much loss to heat.

Measures of Diversity

The word diversity makes some people roll their eyes, dismissive of it as a "politically correct" concept with no practical, real-world applications. Yet, in the biological world, which is really the only world we've got, diversity is critical. We can talk about the diversity of life in several ways: in terms of genes, in terms of species, or in terms of ecosystems. No matter how we measure it, the more of it we have, usually the better off things are.

> **Bio Bits**
>
> The need to maintain genetic diversity is so real that researchers worldwide have collaborated on a specially maintained, closely protected seed bank of wild seeds for crop and other plants in the event of a threat to the food supply. This bank, informally known as the "Doomsday Vault," is sited in Norway inside the Arctic Circle.

Species diversity in an ecosystem is critical to the checks and balances of that system. As we reduce this diversity—as we have done on the American prairies, for example—we run the risk of pulling out the lynchpin of the ecosystem and destroying it altogether. And if one ecosystem fails, the effects could reverberate throughout the biosphere because ecosystems, like species, never exist as islands alone.

Species Richness

In this chapter, I've made reference to types of biomes that are "rich in species." In biology as in society, we tend to view diversity as beneficial. But how do we determine how diverse a community of species or an ecosystem is? One consideration is simply assessing how many different species there are. But another consideration is the relative abundance of each species.

Take two trees in a tropical forest. They both provide a home to 10 species of ants. But on one tree, a single ant species makes up 90 percent of the individuals living on the tree, while on the other tree, each of the 10 ant species has equal representation. Which one of the trees would you describe as being more diverse? In this case, it's not just the number of species but how strongly each species is represented.

Extinction: Diversity Lost

While measuring species richness can be tricky, it's pretty easy to nail down what reduces it: *extinction*. Are we in the middle of a sixth mass extinction event? No one has demonstrated that unequivocally. What we have demonstrated, however, are huge losses across ecosystems and, in some cases, enormous losses within a single taxon, like the amphibia. Many of these losses trace straight to human activity, willful or otherwise, and they translate into effects that reverberate in our lives in ways we have yet to fully comprehend.

def•i•ni•tion

When the last representative of a species dies, the species has entered **extinction**, from which there is currently no return. Once lost, lost forever.

This book has covered a lot of territory, from atoms to the biosphere, but if there is one final lesson that I'd like any reader to take home, it is this: extinction is diversity lost. It is a cancer cure vanished, a new fuel source disappeared, a famine assured, an opportunity missed. Something to remember as we continue with our studies of life, past, present, and future.

The Least You Need to Know

- Nutrients, including carbon, nitrogen, water, and phosphorus, cycle, using organisms as an intermediate.

- Biomes develop their character as a result of the interaction of precipitation and temperature.

- Water biomes are delineated based on their location, depth, and availability to light.

- The food chain/food web represents the transfer of energy through an ecosystem, beginning with autotrophs.

- Diversity is critical to ecosystem/biosphere health and can be assessed from different perspectives.

- Extinction occurs with the death of the last representative of a taxon and is currently permanent.

Appendix A

Glossary

action potential   The all-or-nothing depolarization response of a neuron that occurs when a threshold of depolarization is reached and sodium channels open.

activation energy   The energy barrier or "hump" that must be overcome before a biochemical reaction can move forward.

active transport   Energy-requiring transport of molecules against their concentration gradient, from low to high.

adenosine triphosphate   Usually known as ATP. An adenine nucleotide with three phosphates bound to it; the energy-carrying molecule of most cells.

adhesion   The outcome of the interaction of water molecules with non-water molecules, such as a surface or lining.

afferent   Carrying signals from the outside in.

alcohol fermentation   An anaerobic process of cellular respiration that yields ethanol and carbon dioxide as by-products.

alleles   Different forms of a gene.

amino acid   The monomer or building block of a protein. Consists of a carboxylic acid joined to an amino group by a central carbon that also bears a variable group that defines the specific amino acid.

angiosperm   A vascular plant that produces "dressed" seeds and flowers.

antibodies   Proteins produced by B cells. Used to tag infected cells or invaders for destruction; also called immunoglobulins. Consist of two heavy chains and two light chains that form the characteristic Y shape of a single antibody.

apical meristem   The embryonic tissue of the plant at its tips that drives lengthening growth.

apoptosis   Programmed cell death; also called cell suicide.

Archaea   One of the three domains and one of the two prokaryotic domains, the other being Bacteria.

autocrine   Signaling of a cell to itself.

autopolyploidy   The process of increasing the number of sets of chromosomes through self-fertilization; common to plants.

autosome   Chromosome that is not immediately involved in sex determination. In humans, all chromosomes except X and Y.

Bacteria   One of the three domains of life and one of the two prokaryotic domains; the other is Archaea.

Batesian mimicry   A type of mimicry in which a harmless species mimics a harmful one so that the harmless species benefits.

binary fission   Division in prokaryotes, single-celled eukaryotes, and some organelles, such as mitochondria.

Calvin cycle   The cycle of reactions during the light-independent (dark) reactions of photosynthesis.

carbohydrate   A class of the four big biomolecules; a polar molecule consisting of carbon, hydrogen, and oxygen, also called sugar.

cellular respiration   The breakdown of organic molecules to harvest their energy for building ATP, using set of redox reactions.

channel proteins   Proteins that span the plasma membrane and form aqueous channels, usually for the passage of charged particles like ions.

chlorophyll   Green pigment that reflects light waves in the green spectrum and absorbs primarily at the red and blue wavelengths; a key pigment in photosynthesis.

chloroplast   A double-membrane-bound organelle where the process of photosynthesis takes place.

cholesterol   A lipid, technically a steroid with an alcohol, occurring in the cell membrane, primarily in animals; also the chemical precursor for steroid hormones and vitamin D, among other molecules.

chromatid   The sister copy of a chromosome, attached to the original at the centromere.

chromatin   The combination of DNA packaged with protein.

chromosome   A discrete packet of DNA packaged together with proteins.

codominance   In inheritance, when a heterozygote expresses the phenotype of each of the alleles in its gene pair for a trait. Ex. AB blood type.

codon   A sequence of three bases that serves as a code designating a specific amino acid or the termination of translation.

coenzyme   A cofactor that aids in enzymatic activity.

commensalism   A symbiotic relationship that benefits one species and has no effect either way.

competitive exclusion   The idea that competition between two very similar species for the same niche will end in the extinction of one unless a compromise is made.

competitive inhibition   A regulatory pathway targeting enzymes in which a molecule inhibits the binding of the enzyme to a substrate by occupying the enzyme's active site.

consumer   Any nonphotosynthesizing organism in a food chain; can be primary (eating primary producers), secondary, tertiary, or quaternary.

convergent evolution   The process by which organisms evolve similar adaptations because of occupying similar ecological niches, even though they may not be closely related and/or occupy different geographical regions.

covalent bond   The result of the sharing of a pair of electrons between two atoms.

crossing over   The exchange of genetic material between homologous chromosomes during prophase of meiosis I.

cytokinesis   Splitting of the cytoplasm, the typical final stage of a cell cycle.

cytoplasm   The internal environment of the cell, encompassing the organelles, cytosol, and other structures.

cytoskeleton   An intracellular structure that serves many purposes in prokaryotes and eukaryotes, including cell structure and shape, motion, and intracellular transport in eukaryotes.

cytosol   The liquid component of the internal environment of the cell, outside of the organelles.

dark reactions   The reactions of photosynthesis that can take place independent of sunlight.

dehydration synthesis   The process of linking monomers or building blocks together through covalent bonds formed by the removal of a water molecule.

deoxyribose   The sugar component of a DNA nucleotide; a pentose sugar.

depolarization   A shift from the negative membrane potential of the inner membrane toward the positive.

diffusion   The passive movement of molecules from high to low concentration.

diploid   Having two sets of chromosomes per somatic cell.

DNA   Deoxyribonucleic acid. A nucleic acid that is used to hold the genetic code for building proteins.

dominant allele   An allele is considered dominant if only one copy of it is required for its related phenotype to manifest.

ectoderm   One of the three primary tissues arising from gastrulation, forming the outermost layer.

efferent   Carrying signals from the inside out.

electronegativity   The ability of an atom to attract electrons.

endergonic   A reaction involving the input of free energy (energy in).

endocrine   Related to a system of communication involving signaling molecules called hormones.

endocytosis   An active process by which the cell takes in molecules from the external environment.

endoderm   One of the three primary tissues arising from gastrulation, forming the innermost layer.

endoplasmic reticulum   A network of interconnected tubules, vesicles, and sacs.

endosperm   Triploid tissue in angiosperms formed from the fusion of a pollen nucleus and diploid central cell; nourishes the plant embryo.

entropy   The measure of the disorder of a system.

enzyme   A molecule, usually a protein, that acts as a catalyst for biochemical reactions by lowering the activation energy required for the reaction to move forward.

epidermis   In plants, a tissue of the dermal tissue system derived from the protoderm of the apical meristem.

epistasis   The interaction of two or more independent gene pairs to determine a phenotype.

epithelium   Tissue formed by cells tightly packed together in sheets.

equilibrium   In biology, usually related to equal molecular concentrations on either side of a membrane.

euchromatin   DNA and its proteins loosened up and available to be transcribed.

eudicot   A plant classification based on the presence of two seed leaves; was once called dicot.

Eukarya   One of the three domains of life; the domain that encompasses all eukaryotic organisms.

eukaryote   A type of cell that contains a nucleus and membrane-bound organelles; the cell type that makes up all life assigned to the Eukarya domain and excludes organisms in the Bacteria and Archaea domains, which are prokaryotes (lacking a nucleus). Adj. eukaryotic, meaning to have a nucleus.

exergonic   A reaction that releases free energy (energy out).

exocrine gland   A gland that secretes its contents into ducts.

exocytosis   The active process by which the cell releases molecules, usually from intracellular vesicles, into the external environment.

exons   Coding sequences of DNA.

facultative   Refers to organisms that have a choice in terms of biochemical processes, usually related to metabolism.

fatty acid   A hydrocarbon chain attached to a carboxylic acid; the major component of a triglyceride.

fermentation   The anaerobic pathways of cellular respiration.

filtrate   The fluid that the excretory system extracts or filters from the body.

fitness   Related to the proportion of an individual's genes that occur in a population; often discussed in terms of reproductive success.

fixed action pattern   An innate behavior programmed in the genes that does not change with experience.

fluid mosaic   The current model of the plasma membrane, considering it as a fluid environment of diverse molecules.

founder effect   The effect in a population of reduced genetic variation if the population traces to only a few founding individuals.

frameshift mutation   A mutation that results in a shift of the reading frame.

free energy   The energy available to do work on a system.

functional groups   A group of atoms that confers specific characteristics and behaviors (functions) on a molecule.

fundamental niche   The niche a species might occupy in the absence of competition from similar species.

gamete   The result of meiosis, cells that have half the chromosome number of the parent cell. Ex. sperm or egg.

gametophyte   The multicellular structure that arises from a haploid cell in organisms that exhibit alternation of generations.

gap junctions   Protein channels through cell membranes that allow cytoplasmic communication between cells for passage of materials.

gastrulation   The embryonic process that results in the formation of the initial three tissue layers of ectoderm, mesoderm, and endoderm.

gene expression   The process of transcribing a gene and using its code to build a protein (translation).

gene flow   The movement of alleles into or out of a population as a result of migration.

genetic drift   The process by which an allele becomes either fixed or lost in a population because of an absence of any selection pressure related to it.

genome   A complete set of DNA, e.g., all nuclear DNA in a cell.

genotype   The genetic combination that underlies a phenotype.

genus   The classification category just above the species level and the first of the two terms we use to designate a species; e.g., in *Homo sapiens*, *Homo* is the genus.

glucose   A carbohydrate and the building block (monomer) of starch, glycogen, and cellulose.

glycogen   The branched polymer of glucose molecules used for sugar storage in animals.

glycolysis   The breakdown of glucose into two molecules of pyruvate (pyruvic acid).

glycoprotein   A protein with a branched carbohydrate polymer attached to it, often emerging through the membrane to the outside and used for chemical signaling.

Golgi apparatus   An organelle consisting of membrane-bound stacks called cisternae. Responsible for packaging and modifying proteins and other biomolecules.

grana   Stacks of thylakoids in a chloroplast. Sing. granum.

growth factor   Usually a cellular signaling molecule that triggers a growth response in the form of cell division.

gymnosperm   A vascular plant that makes "naked" seeds, i.e., no fruit or other seed accouterments. Ex. pine.

haploid   Having one set of chromosomes in a cell, or having half of the organism's somatic chromosome number.

heterochromatin   Tightly packed DNA that is not available for transcription.

heterozygous   Describing an individual (heterozygote) carrying different alleles in its gene pair for a trait.

homeostasis   Regulation of balance in the internal environment.

homologous pair   A pair consisting of a paternal and a maternal chromosome with genes in the same order on each chromosome.

homologous trait   A trait that taxa share because of a common ancestry.

homozygous   Describing an individual (homozygote) carrying identical alleles in its gene pair for a trait.

homozygous dominant   The condition in which the individual carries two identical, dominant alleles in its gene pair for a trait.

homozygous recessive   The condition in which the individual carries two identical, recessive alleles in its gene pair for a trait.

hormone   A blood-borne signaling molecule, usually either of lipid or amino acid/protein origin.

hydrolysis   The process of breaking covalent bonds holding a polymer together through the addition of a water molecule.

hydrophilic   Literally, "water loving"; a feature of molecules with a charge or that are ions. These molecules usually interact with water or other polar or charged molecules and are typically soluble in water but not in fat.

hydrophobic   Literally, "water fearing"; a feature of molecules that lack charge and generally do not interact with polar or charged molecules, including water. Not soluble in water but may be fat soluble.

intermolecular bonds   Bonds formed between molecules, such as a hydrogen bond.

interphase   The longest phase of the cell life cycle, consisting of three stages: G1, S, and G2.

intramolecular bonds   Bonds formed within a molecule, such as covalent or ionic bonds.

introns   Noncoding sequences of DNA in eukaryotic genes.

keystone species   A species whose relevance to the ecosystem is out of proportion with its representation.

kinase   A class of enzymes with the job of adding a phosphate onto a molecule.

kinetic energy   Energy of motion.

kinetochore   Region of the centromere where microtubules attach in mitosis to pull sister chromatids apart.

Krebs cycle   The first aerobic step of cellular respiration in which energy from organic molecules entering the cycle is transferred to electron carriers for transport to the electron transport chain; takes place in the mitochondrial matrix.

lactic acid fermentation   An anaerobic pathway of cellular respiration that yields lactic acid as a by-product.

lateral meristem   The embryonic plant tissue responsible for thickening or widening growth of the plant.

ligand   A molecule that interacts with the binding site of a protein.

lipid   A class of the four big biomolecules; includes dietary fats, cholesterol, and phospholipids.

lysogenic cycle   The "benign" infection of a virus in a bacterium, in which the viral genome is silently incorporated into the bacterial genome and replicates with it.

lysosome   A membrane-bound organelle that breaks down macromolecules through hydrolysis.

lytic cycle   The deadly form of infection of a virus in a bacterium in which the viruses hijack the bacterial machinery to produce more viruses, eventually destroying the host.

macrophage   A type of white blood cell that phagocytoses infected cells and other things.

meiosis   Specialized cell division that halves the chromosome number for sexual reproduction.

meristem   A source of embryonic tissue in the plant that allows it to continue growth throughout its life.

mesenchyme   A tissue formed by cells in loose conformation.

mesoderm   One of the three primary tissues arising from gastrulation, forming the middle tissue layer of the embryo.

mesophyll   Cells in the leaf specialized for photosynthesis.

metabolism   All of the biochemical processes that take place in an organism.

microfilaments   In eukaryotes, thin filaments made of actin that form part of the cytoskeleton.

microtubules   The thick filaments of the cytoskeleton in eukaryotes, made of tubulin proteins. Involved in intracellular transport and cell division.

mitochondria   Cellular organelles where most of cellular respiration and the building of ATP takes place. A double-membrane-bound organelle with its own DNA and ribosomes. Sing. mitochondrion.

mitosis   A stage of cell division; specifically, division of the nucleus.

monomer   A unit or building block for large molecules. Ex. nucleotides used to build DNA.

monosomy   Having only one chromosome, instead of two, in one of the pairs of the chromosome set.

mRNA   Messenger RNA, the clean copy of the genetic code that enters the cytoplasm for translation. The result of transcription and other processes in the nucleus.

Müllerian mimicry   A type of mimicry in which a harmful species mimics another harmful species, with the result that both species may benefit.

multiple alleles   In humans, when a trait has more than two possible alleles encoding it.

mutualism   A symbiotic relationship between species that is mutually beneficial.

neurulation   The process in early embryonic development that results in production of the brain and spinal cord.

niche   How an organism/species fits into and uses its environment.

nonpolar   Having no areas of charge. Usually, nonpolar molecules are also hydrophobic.

nonpolar covalent bond   A covalent bond formed between atoms of roughly equal electronegativities.

nucleic acid   A class of the four big biomolecules; the two representatives of this class, which consist of nucleotides linked in a chain, are deoxyribonucleic acid (DNA) and ribonucleic acid (RNA).

nucleolus   Site of ribosomal RNA synthesis in the nucleus.

nucleosome   A level of DNA packing, the "bead" formed by DNA wrapped around histones.

nucleotide   The building block or monomer of a nucleic acid; for DNA or RNA, each nucleotide consists of a sugar (ribose for RNA; deoxyribose for DNA), a phosphate, and a base (AGCT for DNA; AGCU for RNA).

nucleus   Membrane-bound organelle found in many eukaryotic cells; houses the DNA and related proteins and the site of the nucleolus, where ribosomal RNA is transcribed.

obligate   Refers to organisms that must use a specific pathway for metabolism.

Okazaki fragments   Short DNA sequences on the lagging strand in replication, separated by gaps where RNA primers were.

oogenesis   Meiosis resulting in the production of an egg.

origin of replication   The meeting point of two replication forks, where replication in each direction begins.

osmosis   A special case of diffusion, referring only to the movement of water molecules from high to low concentration, across a semipermeable membrane.

outgroup   In phylogenetics, the species or taxon that represents the ancestral condition, having no shared, derived traits relative to the other taxa under consideration.

oxidize   When a molecule has an electron removed.

paracrine signaling   Signaling between neighboring cells.

parasitism   A symbiotic relationship between two species in which the host suffers harm and the parasite benefits.

parenchyma   In plants, a thin-walled tissue arising from the ground tissue system and forming many of the plant's soft or spongy tissues.

passive   Related to movement of molecules, their movement from a high to low concentration, not requiring an energy input. An exergonic reaction that releases energy.

peptide   A short sequence of amino acids.

peptide hormone   A type of blood-borne signaling molecule that consists of amino acids.

phagocytosis   The ingestion of solid particles by the cell, a process requiring energy input.

phenotype   An observable characteristic of an organism.

phosphatase   An enzyme that removes a phosphate from a molecule.

phospholipid   A kind of lipid and the major component of cell membranes; consists of a polar head that interacts well with water and two fatty acid tails that are hydrophobic. Forms a bilayer in the cell membrane with the tails of each layer oriented inward.

phosphorylation   The process of adding a phosphate to a molecule.

phototrophs   Organisms, such as bacteria, that use sunlight as their energy source.

phylogenetics   The study of biological evolutionary history.

pinocytosis   Known more familiarly as "cell drinking," the active process by which the cell takes in extracellular fluid and its solid components.

plasma membrane   The phospholipid bilayer that separates the internal world of the cell from the external world. Also contains proteins, cholesterol, and other molecules and exists as a fluid mosaic that serves as the cell's first line of security.

plasmodesmata   The plant's version of a gap junction: protein channels allowing exchange of materials between plant cells.

polar   Having distinct ends or parts, as in a water molecule. Polar molecules are usually hydrophilic.

polar covalent bond   A covalent bond formed between atoms of unequal electronegativities, resulting in a polar molecule.

pollen   The plant male gametophyte that harbors the male gametes.

pollen tube   Extension from pollen into female structure in plants.

polymer   A large molecule built from smaller units or building blocks referred to as monomers. Ex. DNA built from nucleotides.

polymerase   Enzyme that builds polymers or chains from building blocks (monomers).

population   A group of organisms that are all members of the same species, often occupying the same geographical area. The smallest biological unit that evolves.

primary producer   The base of the food chain, the first trophic level, consisting of organisms that photosynthesize. Contains the most biomass of any tropic level in the chain.

primary succession   The formation of a community beginning with bare rock; usually occurs over long periods of time.

primitive trait   A character that is/was present in the common ancestor of a group of organisms.

prokaryote   A single-celled organism lacking a nucleus. Bacteria and Archaea are all prokaryotes. Adj. prokaryotic; refers to cell type lacking nucleus.

promoter   A sequence in the DNA that allows binding of the machinery that initiates transcription. A major player in regulation of gene expression.

protease   A protein that breaks down other proteins.

protein   A class of the four big biomolecules; proteins consist of amino acids linked together in a chain.

proton   A positively charged particle in the nucleus of an atom; also, a hydrogen ion (H+).

pyruvate   The three-carbon molecule resulting from the process of glycolysis; also called pyruvic acid.

recessive allele   An allele is considered recessive if two copies are required for the related phenotype to manifest.

redox reaction   The combination of oxidation (taking away electrons) and reduction (adding electrons).

reduced   When a molecule has an electron added to it.

replication   The copying of DNA into DNA.

replication bubble   The conformation that results from two replication forks facing each other at their mutual origin of replication.

replication fork   Forklike formation in which each of the two "tines" are parent DNA strands being used as templates for new DNA strands; replication is occurring in the direction where the "tines" intersect to form a *V*.

resource partitioning   The divvying up of a niche as a result of competition between similar species for the same niche.

ribose   The sugar component of an RNA nucleotide; a pentose sugar.

ribosome   A cellular structure found in the cytoplasm and consisting of RNA and associated proteins assembled into two subunits. Responsible for "reading" the instructions (messenger RNA) for building proteins and the site of protein synthesis. Can occur freely or bound to endoplasmic reticulum.

ribozyme   A form of RNA that behaves as an enzyme; under certain conditions, an RNA ribozyme can catalyze copying of RNA.

RNA   Ribonucleic acid. The nucleic acid that has three separate major functions in the cell involved with using the genetic code to build proteins.

rough endoplasmic reticulum   Consisting of the endoplasmic reticulum and associate ribosomes; the site of protein synthesis.

sarcoplasmic reticulum   A kind of smooth endoplasmic reticulum found in muscle cells; responsible for calcium storage and release that triggers muscle contraction.

sclerenchyma   A plant tissue that forms part of the ground tissue system and is characterized by consisting of dead cells at maturity. Provides support and storage.

secondary growth   In plants, the widening growth driven by the lateral meristem.

secondary succession   The formation of a community starting with soil; usually occurs over a shorter period of time.

self-cross   In genetics, using a plant's own pollen to fertilize its ovule.

semiconservative   Refers to the nature of DNA replication; the result of the process conserves one of the parental strands in each new double strand, so the parental strands are semiconserved.

smooth endoplasmic reticulum   A part of the endoplasmic reticulum involved in lipid synthesis and detoxification.

somatic cell   In multicellular organisms, the cells that do not undergo meiosis or are products of meiosis.

speciation   The process by which new species arise from existing species.

spermatogenesis   Meiosis that results in the production of sperm.

spliceosome   Collection of nuclear molecules that removes introns from newly transcribed RNA.

sporophyte   In organisms exhibiting alternation of generations, the multicellular structure that arises from a diploid fertilized cell, destined to produce haploid spores.

starch   A branched polymer of glucose molecules, all oriented in the same way, used for energy storage in plants.

start codon   In translation, the codon that alerts ribosomes to begin the process; always AUG, which is code for the amino acid methionine.

steroid hormone   A kind of lipid employed as a signaling molecule; includes testosterone, estrogen, and progesterone.

stomata   Pores in the leaf and stem for gas exchange. Sing. stoma.

stop codon   Three codons that trigger termination of translation: UAG, UGA, UAA.

stroma   In chloroplasts, the fluid that surrounds the internal structures of the organelle.

symbiosis   Two species living together.

sympatric   Refers to populations occupying the same geographical area; often used to describe a mechanism of speciation.

symplesiomorphy   A shared, primitive or ancestral trait, not so informative for distinguishing evolutionary relationships among taxa.

synapomorphy   A shared, derived trait; useful in determining evolutionary relationships among taxa.

synapsis   The physical pairing of homologous chromosome pairs during prophase of meiosis I, as part of the process of crossing over.

taxon   A group of organisms considered as a related unit, such as a species or all the species in a genus. Pl. taxa.

thylakoid   A membrane-bound compartment in the chloroplasts of plants and in some other photosynthesizing organisms where the light reactions of photosynthesis take place.

tight junctions   Junctions joining cells together so tightly that liquid cannot penetrate.

transcription   The process of copying the genetic code from DNA into RNA.

transcription factor   Proteins, usually in the nucleus, that respond to chemical messages by regulating gene transcription.

transduction (signal)   The process cells use to convert signals from one form to another, usually using biochemical pathways.

transformation   Change in genotype or phenotype resulting from uptake of DNA; common in prokaryotes.

translation   The process of using the code embedded in messenger RNA to build proteins.

transport protein   A protein that spans the plasma membrane and plays a role in the transport of molecules across the membrane.

triglyceride   A dietary fat consisting of a glycerol head with three fatty acid tails attached. Can be saturated or unsaturated.

trisomy   Having three chromosomes, instead of a pair, for one of the chromosome pairs in a chromosome set.

tRNA   Transfer RNA, the nucleic acid that carries the anticodon sequence associated with a specific amino acid. The anticodon sequence, if matched to a codon during translation, results in the tRNA transferring the amino acid it carries to the growing peptide chain.

trophic level   Feeding or energy level in a food chain.

vacuole   A large storage vesicle, usually for storing water (as in a plant's central vacuole) or molecules suspended in water.

vascular   Having a vessel system for nutrient and fluid transport; a way of classifying plants or plant tissues.

vesicle   A small membrane-enclosed sac in the cell used for transporting molecules.

X-linked inheritance   Refers to phenotypes traceable to genes on the X chromosome.

xylem   The tubes in a plant that carry water.

zygote   The cell that results from the union of two gametes, typically with the somatic chromosome number restored.

Readings and Web Resources

Readings

Campbell, Neil, et al. *Biology*, 8th ed. San Francisco: Benjamin Cummings, 2007.

The standard biology text for majors continues to be Campbell et al., *Biology*, in its eighth edition as of this writing. This comprehensive text, covers it all in 1,267 pages, not including appendixes, and weighing in at a whopping 7.5 pounds, by my scale. I first acquired this text when it was in its third edition, and it remains a standby to this day.

Futuyma, Douglas J. *Evolution*, 2nd ed. Sunderland, MA: Sinauer Associates, 2009.

Futuyma, Douglas J. *Evolutionary Biology*, 3rd ed. Sunderland, MA: Sinauer Associates, 1997.

The key texts on evolution are Douglas J. Futuyma's *Evolution* and its denser predecessor, *Evolutionary Biology*.

Web Resources

General Biology

First, check out collegebio.net, the website specific to this book. Collegebio.net provides a chapter on microbes with links, an expanded and comprehensive glossary, updates, errata, a blog for biology-related Q&A with questions answered by Yours Truly, and more materials related to college biology. Be sure to check it out.

Access Excellence: Okay, it's for teachers, but you're obviously the self-teaching type, or you wouldn't be reading this book. Be sure to check out The Living Skeleton link: www.accessexcellence.org

The Biology Project, the University of Arizona: www.biology.arizona.edu/cell_bio/ cell_bio.html

Teacher's Domain offers activities for grades K through 12, broken down by topic within biology and labeled with the appropriate grade level. Many of these are interactive or videos: www.teachersdomain.org/collection/k12/sci.life

Continuing in the self-teaching theme, you can also find educational resources at BioEd Online, produced by the Baylor College of Medicine: www.bioedonline.org

Find anything and everything biology at Kimball's biology pages, the brainchild of John Kimball at Harvard: http://biology-pages.info

The University of California, San Diego, brings you life science information via its television-show website, Science Matters. Because it does: www.ucsd.tv/sciencematters

The Howard Hughes Medical Institute brings you BioInteractive, with video, animation, virtual labs, and other tools that "teach ahead of the textbook": www.hhmi.org/ biointeractive

Cells

Cells Alive! Find all things cell, including animations and videos: www.cellsalive.com

What is a cell? The National Center for Biotechnology Information can tell you: www.ncbi.nlm.nih.gov/About/primer/genetics_cell.html

The University of Arizona's Biology Project, an online learning tool, has a section specific to cells: www.biology.arizona.edu/cell_bio/cell_bio.html

Wiley offers these interactive animations of cells in the context of one of its biochemistry texts; you can even build your own cell: www.wiley.com/college/boyer/0470003790/animations/cell_structure/cell_structure.htm

Enjoy a color-rich tour of the cell, courtesy of the U.S. National Science Foundation: www.nsf.gov/news/overviews/biology/interactive.jsp

An 8-minute animation of the inner life of a cell. This is just flat-out cool, and I show it to my students frequently: www.studiodaily.com/main/technique/tprojects/6850.html

Genetics

PBS has an online DNA workshop with interactives and information about the molecule and the people involved with DNA: www.pbs.org/wgbh/aso/tryit/dna/index.html

Teacher's Domain, which may require free registration, has an abundance of videos and other useful information. Here are a couple of links to interactive genetics activities, including one on Gregor Mendel: www.teachersdomain.org/asset/hew06_int_dominantgene; www.teachersdomain.org/asset/hew06_int_mendelinherit

Learn what we've learned about the human genome at the Human Genome Project website: www.ornl.gov/sci/techresources/Human_Genome/home.shtml

The website of the Genetic Science Learning Center at the University of Utah pretty much has it all, from gene expression to epigenetics. A beautiful site with a clear and visual presentation of information: learn.genetics.utah.edu/

A nice breakdown of a great two-hour *NOVA* episode on genetics into digestible video segments. Just pick your topic: www.pbs.org/wgbh/nova/genome/program.html

Evolution

Gain access to the original works of the man himself, Charles Darwin, at The Darwin Digital Library website, based at the American Museum of Natural History library: darwinlibrary.amnh.org

This site from the University of California Museum of Paleontology and the National Center for Science Education bills itself as a "one-stop source for evolution," so I guess I could stop here with providing evolution links (but I won't): evolution.berkeley.edu/evolibrary/home.php

The University of California Museum of Paleontology also has an amazing website that covers aspects of evolution, taxonomy/systematics/diversity, the history of science, and, of course, fossils. Absolutely worth checking out: www.ucmp.berkeley.edu/help/topic.html

Macroevolution.net offers evolution information, including specifics on human evolution, and lots of links to more: www.macroevolution.net/biology-websites.html

Explore the creation/evolution controversy over at Talk Origins, the web's most comprehensive and scientific treatment of the topic: www.talkorigins.org

Diversity

The best place to go for an understanding of the diversity of life on Earth and the relatedness of species, the Tree of Life Web Project: tolweb.org

Oh, yes. Check out the University of Michigan's Animal Diversity Web and browse the kingdom: animaldiversity.ummz.umich.edu/site/index.html

Absolutely relevant to issues of diversity are endangered species. Find out all about it here: eelink.net/EndSpp

I think "Diversity" is a good place to put the link to Action Bioscience, which addresses current knotty issues in biology: www.actionbioscience.org

The Encyclopedia of Life, the brainchild of E. O. Wilson, seeks to document in a free, online resource, each of the 1.8 million species we've currently identified. Pretty pictures and good information, so check it out: eol.org

Because fossils are part of the earth's diversity, I also bring you the Paleobiology Database: paleodb.org/cgi-bin/bridge.pl

Plants

Enjoy a comprehensive site from the Botanical Society of America, including an ever-popular "Botany in the News" department: www.botany.org

An online hyperlinked botany textbook for people like you, and if you're interested, a good chance to practice your German: www.biologie.uni-hamburg.de/b-online/e00/default.htm

For enthusiasts, I bring you Botany.com, an online encyclopedia of plants and flowers: www.botany.com

For real enthusiasts, I recommend the Smithsonian Institution's Index Nominum Genericorum: botany.si.edu/ing

Learning about plants can be confusing. These two articles by David Hershey from Action Bioscience lay out and clear up a few confusions and misconceptions about plants: www.actionbioscience.org/education/hershey.html; www.actionbioscience.org/education/hershey3.html

This online virtual library of plants and botany is courtesy of the University of Oklahoma: www.ou.edu/cas/botany-micro/www-vl

Body Systems

Gray's Anatomy, it's the real book, not the TV show: www.bartleby.com/107

A virtual anatomy textbook! Learn your systems visually: www.acm.uiuc.edu/sigbio/project/index.html

The Visible Human Project—get a peek into the human body, slice by slice: www.madsci.org/~lynn/VH

Get Body Smart. Really, that's the website. It offers interactive flash animations about body systems, a great resource if you need or want more detail: www.getbodysmart.com

Find more on the senses from the Howard Hughes Medical Institute: www.hhmi.org/senses

Developmental Biology

A website associated with Scott F. Gilbert's excellent developmental biology text, it's essentially a dev bio textbook online: 8e.devbio.com

The Virtual Embryo offers another comprehensive look at developmental biology, including in-depth information on different animal models: people.ucalgary.ca/~browder/virtualembryo/db_tutorial.html

The Visible Embryo is pretty much self-explanatory—a tour of human development: www.visembryo.com/baby

Developmental biology studies rely heavily on a specific set of animal models. Learn more about those models here: www.ceolas.org/VL/mo

No learning experience in developmental biology is complete without real movies of real development. Find some here at Dynamics of Development 2.0, courtesy of Jeff Hardin at the University of Wisconsin, Madison: worms.zoology.wisc.edu/dd2

Ecology

The Ecological Society of America offers up a comprehensive website of facts, publications, and information about actual ecologists: www.esa.org/education_diversity

Learn more about invasive species and what's being done about them at the Global Invasive Species Program website: www.gisp.org

Not sure about which biome you occupy? Check out the Biomes of the World website, courtesy of the Missouri Botanical Garden, and find out: www.mbgnet.net

Ecosystems present and future, watch this series brought to you by PBS and Bill Moyers, in association with the *Earth on Edge* series: www.pbs.org/earthonedge/ecosystems/index.html

Get the global skinny on ecosystem health and prognosis from the United Nations and partners in the Millennium Ecosystem Assessment: www.millenniumassessment.org/en/index.aspx

Index